# WEB THEORY
## AND RELATED TOPICS

# WEB THEORY
## AND RELATED TOPICS

edited by

**Joseph Grifone**
*Université Paul Sabatier*
*France*

**Eliane Salem**
*Université Pierre et Marie Curie*
*France*

**World Scientific**
*Singapore • New Jersey • London • Hong Kong*

*Published by*

World Scientific Publishing Co. Pte. Ltd.

P O Box 128, Farrer Road, Singapore 912805

*USA office:* Suite 1B, 1060 Main Street, River Edge, NJ 07661

*UK office:* 57 Shelton Street, Covent Garden, London WC2H 9HE

**British Library Cataloguing-in-Publication Data**
A catalogue record for this book is available from the British Library.

**WEB THEORY AND RELATED TOPICS**

ISBN 981-02-4604-8

Printed in Singapore by Uto-Print

0013009547

# PREFACE

This volume contains the contributions to the *Journées sur les Tissus*, held in Toulouse in December 1996, thanks to a visit of Prof. I. Nakai.

This meeting was organized by the Laboratoire Emile Picard, Université Paul Sabatier, together with the Ecole Doctorale, and the European Network "Singularités des Equations Différentielles et Feuilletages".

We thank these institutions for their support. We also express our gratitude to the participants and particularly to the authors.

Toulouse, December 2000
*J. Grifone and E. Salem*

Joseph GRIFONE
Laboratoire Emile Picard,
UMR CNRS 5580
Université Paul Sabatier
118 route de Narbonne
F31062 Toulouse, Cedex
grifone@picard.ups-tlse.fr

Eliane SALEM
Institut de Mathematiques,
UMR CNRS 9994
Universite de Paris 6,
46-56, 5 eme etage, boite 247
4 place Jussieu
F75252 Paris, Cedex 05
salem@math.jussieu.fr

v

# CONTENTS

# CONTENTS

# A NAIVE GUIDE TO WEB GEOMETRY

ISAO NAKAI

*Department of Mathematics, Faculty of Science, Ochanumizu University,*
*Bunkyo-ku, Ontsuka 2-1-1 Tokyo 112-8610, Japan, nakai@math.ocha.ac.jp*

Web structure, in other words a configuration of foliations can be seen often in the mathematical nature. For example $d$ pencils of lines in the plane passing through $d$ base points exhibit a $d$-web structure of codimension 1. This structure is classically known to be hexagonal: any 3-subweb consisting of 3 pencils is locally diffeomorphic to the 3-web by parallel lines [2,3,11]. Many theorems, such as Pappus' theorem, have been stated on this structure. A configuration of $d$ lines in the dual projective plane corresponds to these $d$ pencils. In general for a projective curve $C \subset \mathbb{P}^n$ of degree $d$ the dual hyperplanes of those points on $C$ exhibit a $d$-web structure of codimension one on the dual projective space [11,22]. Pascal's theorem, Reiss' theorem, Graf-Sauer's theorem and the group structure on cubic curves explain the hexagonality of the 3-web on the dual projective plane [2,3,11].

One of the most meaningful results in web geometry is Lie's theorem found at the end of the 19 th century [26,27,28]. A germ of a translation hypersurface $S \subset \mathbb{C}^n$ is a surface presented as $S = C_1 + \cdots + C_{n-1}$ with germs of space curves $C_i$. The surface $S$ is called a double translation surface if it admits another translation structure. By definition a $(2n-2)$-web structure of codimension 1 is seen on those surfaces. It is classically known that the theta divisor in the Jacobian $J(C)$ of a Riemann surface $C$ is naturally a double translation surface by Abel's theorem [22]. Now let $C \subset \mathbb{P}^{n-1}$ be a canonical curve of degree $2n - 2$. The dual hyperplanes of those points on $C$ exhibit a $(2n - 2)$-web structure on the projective dual space. The theta divisor is a covering of the dual space branched along the dual hypersurface of $C$. The pull back of the $(2n - 2)$-web to the theta divisor is nothing but the web given by the double translation structure. Lie's theorem was immediately generalized by Poincaré and later by Darboux, Griffiths and Henkin [14,22,24,31]. The generalized theorem asserts that a germ of a double translation surface is a germ of a $(2n - 2)$-web defined by a projective algebraic curve of degree $2n - 2$ in $\mathbb{P}^{n-1}$, hence it extends to a global analytic hypersurface. If the curve is irreducible, it is a canonical curve. But the curve can degenerate to a union of two rational normal curves of degree $n - 1$ in $\mathbb{P}^{n-1}$. Cayley surface is known to be one of those surfaces. The classification of those degenerate double translation structures seems non complete. Also singularities of double

1

translation structures are not well investigated. The classical web geometry aims at generalizing Lie's theorem to the various cases.

A modern study of web structures from the view point of differential geometry was initiated by E.Cartan in 1908 [7]. He defined, given a differential equation $y' = f(x, y)$, an affine connection on the $xy$-plane which respects the 3 line configuration of the horizontal, vertical lines and the tangent lines of the solutions. This opened a way in a new direction which was followed by people of the Hamburg school, such as Blaschke, Bol and Chern, and after the war mainly by people of the Russian school such as Akivis, Goldberg [1,6,20]. The affine connection defined by Cartan is now called the Chern connection. The author investigated in this volume the singularity of the curvature form of the various singular 3-web structures on the plane defined by implicit differential equations of type $f(x, y, y') = 0$. This has been recently generalized by Mignard [29].

Nagy's paper in this book gives a modern presentation of the Chern connection and curvature theory for 3-webs. For more information on the differential geometry and algebraic investigation on web structures, consult the survey by Akivis and Goldberg [1], and the paper of Henaut in this volume.

An Abelian equation of a germ of a $d$-web of codimension 1, $W = (F_1, \ldots, F_d)$ on the $n$-space is $du_1 + \cdots + du_d$, where $du_i$ is a closed 1-form defining the $i$-th foliation $F_i$ of the web. Blaschke called the dimension of the space of Abelian equations the rank of $W$. For plane 3-webs the rank 1 condition is equivalent to hexagonality. Chern and Griffiths [12] showed in 1970s that the rank is bounded by the Castelnuovo bound for the genus of degree $d$ space curves in $\mathbb{P}^n$. This result was generalized later by many authors (cf. [1]) by analyzing the characteristic variety of a differential system. Henaut [23] found recently a function basically equivalent to the curvature form of Cartan's affine connection by investigating the rank 1 condition for 3-webs with $\mathcal{D}$-module theory.

Another important result in web geometry is Bol's discovery of the exceptional 5-web. Bol proved that all hexagonal $d$-webs on the plane are diffeomorphic to $d$-pencils of lines except for the case $d = 5$. In the case $d = 5$ the Castelnuovo bound is 6. Among 6 Abelian equations he found the five-term relation of Roger's dilogarithm function [4,21]. So far the space of Abelian equations has not yet been well investigated. The exceptional 5-web was generalized by Damiano [13]. From another view point Gelfand and MacPherson [17] investigated the five-term relation.

After a beautiful survey paper by Chern [11] in 1982, new links between web geometry, symplectic geometry and integrable systems were discovered. Brouzet, Molino, Turiel [5] and Fernandes [16] proved that if a flow on a manifold of even dimension admits a bi (= two) Hamiltonian structure, then the Hamiltonian function splits into a sum of functions of one variable near the invariant tori. In particular the graph of the Hamiltonian function is a surface of translation.

It is known that many integrable differential equations such as $KdV$-equation admit bi-Poisson structure with non trivial kernel. So their function spaces are of "infinite-odd dimension" according to Gelfand and Zakharevich [18]. Here a Poisson structure on a manifold $M$ is a 2-vector field on $M$ which satisfies the Jacobi identity. If a Poisson structure has constant kernel rank $c$, the kernel defines a plane field on $M$ which is integrable by the Jacobi identity and defines a symplectic foliation of codimension $c$ on $M$. Gelfand and Zakharevich investigated bi-Poisson structures on a manifold $M$ of finite-odd dimension. By their definition there is a one parameter linear family of bi-Poisson structures on $M$. If each Poisson structure has kernel rank 1, there exists a one parameter family of symplectic foliations on $M$. They showed that the conormal directions of the symplectic foliations are contained in a Veronese space curve in the projectivized cotangent bundle of $M$ at each point, as a consequence of Kronecker's normal form theorem for linear family of skew symmetric matrices. This family of foliations is called the Veronese web. The bi-Poisson flow on $M$ preserves the whole structure. Thus the flow admits transversely Veronese structure. A remarkable observation by Gelfand and Zakharevich is that most integrable differentiable equations are transversely flat. Veronese webs were later studied by Rigal [32] (see also her paper in this volulme). Recently Turiel [34,35] classified local Veronese webs.

A Veronese web is by definition a non linear sub family of a linear family of 1-forms. The linear families of one forms were investigated by and independently Cerveau and Lins Neto [10] and independently by Ghys [19]. Cerveau and LIns Neto showed that the set of projective foliations contains some irreducible components which are unions of projective spaces. For such linear families of one forms, one can define an affine connection. The author characterized webs in such families in terms of the curvature form in [30]. Thom [33], Cerveau [9] and Dufour [15] investigated implicit ordinary differential equations.

Another application of web geometry was made by Joly, Métivier and Rauch [25] for non linear geometric optics. For the details see also the paper by Joly in this volume.

## References

1. M. A. AKIVIS & V. V. GOLDBERG, *Differential geometry of webs*, in

Handbook of differential geometry edited by F.J.E. Dillen, L.C.A. Verstraelen, vol. 1, chapter 1, Elsevier Science B.V., (2000), 1–152.

2. W. BLASCHKE, *Einführung in die Geometrie der Waben*, Birkhäuser-Verlag, Basel-Stuttgart, 1955.

3. W. BLASCHKE & G. BOL, *Geometrie der Gewebe*, Grundlehren der Mathematischen Wissenschaften, Springer, 1938.

4. G. BOL, *Über ein bemerkenswertes 5-Gewebe in der Ebene*, Abh. Math. Sem. Hamburg **11** (1936), 387–393.

5. R. BROUZET, P. MOLINO & F. TURIEL, *Géométrie des systèmes bihamiltoniens*, Indag. Math. **4** (1993), no. 3, 269–296.

6. R. L. BRYANT, S. S. CHERN, R. B. GARDNER, H. L. GOLDSCHMIDT & P. A. GRIFFITHS, *Exterior Differential Systems*, Springer-Verlag, Mathematical Science Reserch Institute Publications,vol. 18, 1991.

7. E. CARTAN, *Les sous-groupes des groupes continus de transformations*, Ann. Ec. Norm. Sup (3) *25* (1908), 57–124.

8. D. CERVEAU, *Equations differentielles algébraiques: remarques et problemes*, J. Fac. Sci. Univ. Tokyo, Sect IA Math. 36 ( 1989), no. 3, 665–680.

9. D. CERVEAU, *Théorème de type Fuchs pour les tissus feuilletés*, Complex analytic methods in dynamical systems (Rio de Janeiro, 1992), Astérisque, no. 222, 1994.

10. D. CERVEAU & A. LINS NETO, *Irreducible components of the space of holomorphic foliations of degree two in* $CP(n)$, $n \geq 3$, Ann. of Math. (2) **143** (1996), no. 3, 577–612.

11. S. S. CHERN, *Web geometry*, Bull A.M.S. **6** (1982), 1–8.

12. S. S. CHERN & P. A. GRIFFITHS, *Abel's theorem and webs*, Jahresber. Deutsch. Math.-Verein. **80** (1978), no. 1-2, 13–110.

13. D. B. DAMIANO, *Webs and characteristic forms of Grassmann manifolds*, Amer. J. Math. **105** (1983), no. 6, 1325–1345.

14. G. DARBOUX, *Lecon sur la théorie générale des surfaces*, Livre I, $2^e$ édition, Gauthier-Villau, Paris, (1914), 151–161.

15. J. P. DUFOUR, *Modules pour les familles de courbes planes*, Ann. Inst. Fourier **39** (1989), no. 1, 225–238.

16. R. L. FERNANDES, *Completely integrable bi-Hamiltonian systems*, J. Dynam. Differential Equations **6** (1994), no. 1, 53–69.

17. I. M. GELFAND, R. D. MACPHERSON, *Geometry in Grassmannians and a generalization of the dilogarithm*, Adv. in Math. **44** (1982), no. 3, 279–312.

18. I. M. GELFAND & I. ZAKHAREVICH, *Webs, Veronese curves and bi-Hamiltonian systems*, J. Funct. Anal. **99** (1991), 150–178.

19. E. GHYS, *Flots transversalement affines et tissus feuilletés*, Mem. Soc. Math. France **46** (1991), 123–150.

20. V. V. GOLDBERG, *Theory of multicodimensional $(n + 1)$-webs*, Kluwer Academic Publ., Mathematics and Its Applications, vol. 44, 1988.

21. V. V. GOLDBERG, *Gerrit Bol (1906-1989) and his Contribution to web geometry*, in Webs and Quasigroups Tver University, Russia (1994).

22. P. A. GRIFFITHS, *Variations on a theorem of Abel*, Invent. Math. **35** (1976), 321–390.

23. A. HÉNAUT, *Sur la courbure de Blaschke et le rang des tissus de $\mathbb{C}^2$*, Natural Science Report of the Ochanomizu University 51 (2000), no. 1, 11–25.

24. G. M. HENKIN, *The Abel-Radon transform and several complex variables*, Modern methods in complex analysis (Princeton, NJ, 1992), Ann. of Math. Stud., vol. 137, Princeton Univ. Press, Princeton, NJ, (1995), 223–275.

25. J. JOLY, G. METIVIER & J. RAUCH, *Caustics for Dissipative Semilinear Oscillations*, Mem. AMS, vol. 114, no. 685, 2000.

26. S. LIE, *Bestimmung aller Flächen, die in mehrfacher Weise durch Translationsbewegung einer Kurve erzeugt werden*, Ges. Abhandlungen. Bd. **1**, 450-467; Arch. für Math. Bd. **7** Heft 2 (1882), 155–176.

27. S. LIE, *Die theorie der Translationflächen und das Abelsche Theorem*, Libzig Ber. **48** (1896), II-III, 141–198.

28. S. LIE, *Das Abelsche Theorem und die Translationsmannigfaltigkeiten*, Leipziger Berichte (1897), 181–248; Ges. Abhandlungen. Bd.II, Teil II, paper XIV, 580-639.

29. G. MIGNARD, *Rang et courbure des 3-tissus de $C^2$*, C. R. Acad. Sci. Paris Sér. I Math. 329 (1999), no. 7, 629–632.

30. I. NAKAI, *Curvature of curvilinear 4-webs and pencils of one forms: variation on a theorem of Poincaré, Mayrhofer and Reidemeister*, Coment. Math. Helv. **73** (1998), no. 2, 177–205.

31. H. POINCARÉ, *Sur les surfaces de translation et les fonctions abéliennes*, Œuvres t.VI, 13-37, Bul. Soc. Math. France 29 (1901), 61–86.

32. M. H. RIGAL, *Systèmes bihamiltoniens en dimension impaire*, Ann. Sci. Ecole Norm. Sup. **31** (1998), no. 3, 345-359.

33. R. THOM, *Sur les équations différentielles multiformes et leurs intégrales singulières*, Bol. Soc. Brasil Mat. **3** (1972), no. 1, 1–11.

34. F. J. TURIEL, *$C^\infty$-équivalence entre tissus de Véronèse et structures bihamiltoniennes*, C. R. Acad. Sci. Paris Ser. I Math. **328** (1999), no. 10, 891–894.

35. F. J. TURIEL, *$C^\infty$-classification des germes de tissus de Véronèse*, C. R. Acad. Sci. Paris Ser. I Math. **329** 1999, no. 5, 425–428.

# ANALYTIC WEB GEOMETRY

ALAIN HÉNAUT

*Laboratoire de Mathématiques pures, Université Bordeaux I et C.N.R.S., F-33405 Talence Cedex, henaut@math.u-bordeaux.fr*

The following text gathers, in an enlarged perspective, the main parts of talks given at the "Journées sur les tissus" conference given at laboratory *Emile Picard* in Toulouse in December 1996. After a short introduction, some problems and results on webs will be presented, beginning with planar webs, before going into generalizations.

## 1  Introduction

Web geometry in $\mathbb{C}^N$ consists in the study of families of foliations of $\mathbb{C}^N$ which are in *general position*. We restrict ourselves to the *local* situation, in the neighborhood of the origin in $\mathbb{C}^N$, of $d$ complex analytical foliations of codimension $n$ in general position. In other words, the leaves of the different foliations share the *same* dimension and their tangent spaces form a family of maximal rank in the ambient space. For instance, in $(\mathbb{C}^2, 0)$ a 3-web (of curves) corresponds to the following picture:

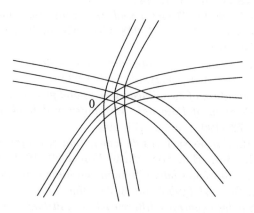

Figure 1.

We are interested in the *geometry* of such configurations, that is properties of $d$-webs of codimension $n$ in $\mathbb{C}^N$ which are *invariant* with respect to

6

analytical local isomorphisms of $\mathbb{C}^N$. As a consequence, we are looking for invariants. These can be either numerical invariants or more geometrical ones: of course local models of webs, canonical forms for leaves, but also curvatures or for instance the projective configurations defined by the normals of the leaves, *etc.*

Projective geometry gives numerous examples of webs which will be used as *standard model*, especially when the codimension $n$ of the leaves is a divisor of the dimension $N$ of the ambient space. This enlightens the title chosen for section 3.

The $C^\infty$ setting seems natural, but the *complex analytic* setting seems more adapted to the study of invariants of webs. For instance, concerning the use of Abel's theorem in order to algebraize a web, or of Chow's theorem, or more generally speaking while using the "GAGA" principle of J.-P. Serre (*i.e.* "global analytic objects of a complex projective algebraic variety are algebraic"). Moreover, one knows that real geometry, even algebraic, is extremely complicated.

In fact, there is another reason for choosing this setting, that is the existence of *singularities* which we avoid for the moment, but whose interest cannot be denied. Effectively, several tools in complex analytic geometry have been developed in order to study singularities and most webs, as everyone can experiment even on simple examples, do have singularities. This is for instance the case of the 3-web $\mathcal{P}(3)$ in $\mathbb{P}^2$ generated by pencils of lines passing through 3 distinct points; the singular locus of $\mathcal{P}(3)$ consists of the points of $\mathbb{P}^2$ where the leaves are singular, or do not intersect transversally, that is the union of the three straight lines passing through two of the given points:

Moreover, the study of singular webs seems of some interest as a tool for *locally generating forms*, as phrased by R. Thom (*cf.* [60]). At least, this should be one goal in the study of *polylogarithmic* webs (*cf.* sect. 3.3).

In the following text, some authors will be thoroughly cited: among them of course the *initiators* of the subject, some of our glorious predecessors, and others. One dares giving the following list, necessarily incomplete, whose sponsorship is at least encouraging, but forces respect.

a) Some glorious mathematicians who, like M. Jourdain wrote in prose, worked on webs without knowing it. In alphabetical order, one can find: N.H. Abel, E. Cartan, G. Castelnuovo, G. Darboux, T.-H. Gronwall, S. Lie, H. Poincaré, M. Reiss, B. and C. Segre, A. Tresse and W. Wirtinger.

b) The initiators of the theory who, in the 30's, got first interested in webs for themselves. Besides W. Blaschke, two other names come to mind imme-

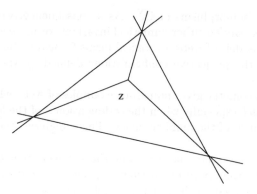

Figure 2.

diately: G. Thomsen and G. Bol. Under the supervision of W. Blaschke, 66 papers have been published between 1927 and 1938 on this subject, under the global title *"Topologische Fragen der Differentialgeometrie"*. Most authors are students or colleagues of W. Blaschke, based in Hamburg. Among the authors, besides the three people already cited, one can find K. Reidemeister, E. Sperner, E. Kähler, O. Zariski, H. Kneser,... and S.S. Chern, whose thesis consists of numbers 60 and 62. The classic book [10] of W. Blaschke and G. Bol entitled *"Geometrie der Gewebe"* gives an overview of numerous results proven in this period (see also [8,9]).

c) Then we have to wait till the 70's for M.A. Akivis and V.V. Goldberg to take the subject to the Soviet Union, joined notably by A.M. Shelekov (*cf.* for instance [2,27,4] and [3]). At the same time, on the other side of the Atlantic, and as soon as the middle of the 70's, P.A. Griffiths gives it a new breath, accompanied by S.S. Chern (*cf.* [32,33,18,19] and [17]); some extrapolations are also due to D.B. Damiano and J.B. Little (*cf.* [20,50] and [51]) in the early 80's. More recently and besides the people attending these "Journées", it is suitable to give the names of D. Cerveau and E. Ghys (*cf.* [15] and [22]) for their contributions.

Though global objects will naturally appear, let us recall that the study is local. One has to state that only few work has been made on *global webs* carried by a differentiable or analytic manifold; however this subject has been recommended by K. Stein in the 60's, and then by F. Norguet as a tool for studying the space of cycles (*cf.* [55]). In order not to make this paper too long, we won't talk about the results concerning webs whose codimension does not divide the dimension of the ambient space, even though this situation gives rise to numerous interesting natural questions.

The next two chapters present some fundamental results of web geometry and also some open questions: some are more or less classical ones, others are related to recent research of the author. Moreover, new results will be given concerning $d$-webs $\mathcal{W}(d, 2, n)$ of codimension $n$ in $(\mathbb{C}^{2n}, 0)$, with indications concerning the proofs. Finally, a substantial bibliography (more than 60 references) is given to ground and help develop the different subjects.

As an ending for this introduction and in order to complete this "Invitation au voyage", here are two examples of planar webs which come from familiar situations:

1. The generic tangents to an algebraic curve $C \subset \mathbb{P}^2$ of class $d$. In the neighborhood of a generic point $z \in \mathbb{P}^2$ we get a $d$-web which is *linear* (*i.e.* whose leaves are straight lines, not necessarily parallel to one-another). For instance when $d = 3$ we get the following picture:

2. The integral curves of a differential equation of the first order and with degree $d$

$$P_0(x,y)(y')^d + P_1(x,y)(y')^{d-1} + \cdots + P_d(x,y) = 0.$$

Near any point $z_0 = (x_0, y_0) \in \mathbb{C}^2$ where both the $y'$-discriminant $\Delta(x_0, y_0)$ and the leading coefficient $P_0(x_0, y_0)$ are non-zero, we get $d$ different tangents and the corresponding integral curves are the $d$ leaves of a $d$-web in $(\mathbb{C}^2, z_0)$:

## 2  Webs in $\mathbb{C}^2$

### 2.1  Abelian relations

Let $\mathcal{O} := \mathbb{C}\{x, y\}$ be the ring of convergent power series in 2 variables. A $d$-web $\mathcal{W}(d)$ in $(\mathbb{C}^2, 0)$ is given by $d$ *families* of curves in general position; these are germs of level sets defined by $F_i(x, y) = cst$, where $F_i \in \mathcal{O}$ can be chosen to satisfy $F_i(0) = 0$. The assumption that the leaves be in general position thus reduces to

$$dF_i(0) \wedge dF_j(0) \neq 0 \qquad \text{for } 1 \leq i < j \leq d.$$

Since we are interested only in the geometry of leaves, the main important objects are the vector fields

$$X_i = \partial_y(F_i)\partial_x - \partial_x(F_i)\partial_y \qquad \text{for } 1 \leq i \leq d$$

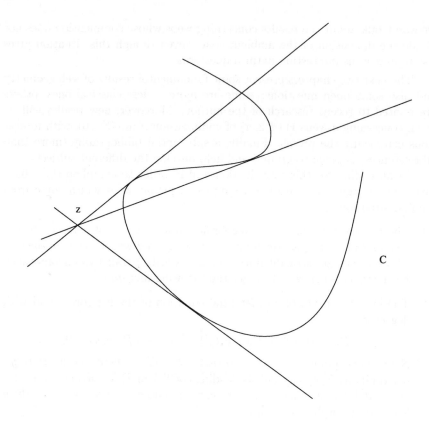

Figure 3.

defined *up to an invertible element of $\mathcal{O}$* or equivalently the corresponding Pfaff forms

$$\omega_i = \partial_x(F_i)dx + \partial_y(F_i)dy \qquad \text{for } 1 \leq i \leq d$$

where $\partial_x = \dfrac{\partial}{\partial x}$ and $\partial_y = \dfrac{\partial}{\partial y}$.

Particularly, when $(x, y)$ is near $0 \in \mathbb{C}^2$, the *normals* $\omega_i(x, y)$ define $d$ different points $[\partial_x(F_i), \partial_y(F_i)]$ in $\mathbb{P}^1$ which depend only on the web $\mathcal{W}(d)$ of $(\mathbb{C}^2, 0)$ and not on the choice of the functions $F_i$.

The main invariant of $\mathcal{W}(d)$ is related to the notion of Abelian relation. A $d$-uple $\big(g_1(F_1), \ldots, g_d(F_d)\big) \in \mathcal{O}^d$ satisfying $\sum_{i=1}^{d} g_i(F_i)dF_i = 0$ with $g_i \in$

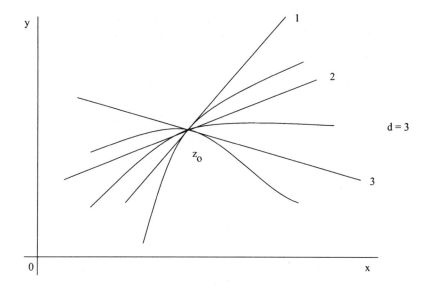

Figure 4.

$\mathbb{C}\{z\}$ is called an *Abelian relation* (of degree 1) of the web $\mathcal{W}(d)$. The $\mathbb{C}$-vector space of Abelian relations of the web $\mathcal{W}(d)$ will be denoted by

$$\mathcal{A}(d) = \big\{ \big(g_i(F_i)\big) \mid g_i \in \mathbb{C}\{z\} \text{ and } \sum_{i=1}^{d} g_i(F_i)dF_i = 0\big\}.$$

**Theorem 2.1** *Let $\mathcal{W}(d)$ be a $d$-web in $(\mathbb{C}^2, 0)$, then the following optimal inequality holds:*

$$\operatorname{rk}\mathcal{W}(d) := \dim_{\mathbb{C}} \mathcal{A}(d) \leq \frac{(d-1)(d-2)}{2}.$$

It is easily checked that the integer $\operatorname{rk}\mathcal{W}(d)$ defined above is an invariant of the web $\mathcal{W}(d)$ which does not depend on the choice of the functions $F_i$. This integer is called the *rank* of the web.

Before we go on and present the ideas underlying several proofs for this inequality and explain the choice for the terminology related to Abelian relations, we have to give the following remarks and proposition. If $d = 1$

(resp. 2), then the local inverse mapping theorem shows that the local model for a web $\mathcal{W}(d)$ is given by the following families of lines in $\mathbb{C}^2$:

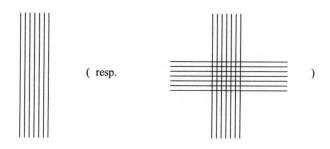

( resp. )

Figure 5.

that is $\{x = cst\}$ (resp. $\{x = cst\}$ and $\{y = cst\}$); this means that the study of possible configurations for the different $\mathcal{W}(d)$ is interesting only for $d \geq 3$.

If $d = 3$, then the previous theorem gives $\operatorname{rk} \mathcal{W}(d) \leq 1$. If $\mathcal{H}$ is a 3-web in $(\mathbb{C}^2, 0)$ with maximal rank 1, then we get a *non-trivial* Abelian relation for $\mathcal{H}$

$$g_1(F_1)dF_1 + g_2(F_2)dF_2 + g_3(F_3)dF_3 = 0.$$

If now we integrate the functions $g_i$, we obtain functions $\tilde{g}_i$ satisfying

$$\tilde{g}_1(F_1) + \tilde{g}_2(F_2) + \tilde{g}_3(F_3) = 0$$

and the general position assumption yields that $\phi = (\tilde{g}_1(F_1), \tilde{g}_2(F_2))$ : $(\mathbb{C}^2, 0) \longrightarrow (\mathbb{C}^2, 0)$ is a local isomorphism which, given the previous relation, carries $\mathcal{H}$ into the linear 3-web whose leaves are $x = cst$, $y = cst$ and $x + y = cst$:

this particular one is of rank 1 since $dx + dy - d(x+y) = 0$ (!). In other words, the *rank* classifies 3-webs in $(\mathbb{C}^2, 0)$ and we get the following proposition:

**Proposition 1** *Let* $\mathcal{W}(3)$ *be a 3-web in* $(\mathbb{C}^2, 0)$, *then* $\operatorname{rk} \mathcal{W}(3) = 1$ *if and only if the 3-web* $\mathcal{W}(3)$ *is defined by* $x = cst$, $y = cst$ *and* $x + y = cst$ *up to an analytical isomorphism of* $(\mathbb{C}^2, 0)$.

At least three different proofs have been given for the previous upper bound of $\operatorname{rk} \mathcal{W}(d)$. One is due to G. Bol, another one to W. Blaschke (following ideas of H. Poincaré about translation surfaces (*cf.* [57])) and a third one

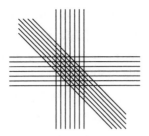

Figure 6.

using basic properties of algebraic analysis (*i.e.* algebraic theory of linear differential systems also called $\mathcal{D}$-modules theory). The first two proofs use the same following method: consider an *independent family* $\left(\gamma_i^j(F_i)\right)_{1\leq i\leq d,\,1\leq j\leq r}$ of Abelian relations for the web $\mathcal{W}(d)$ and differentiate several times the relations

$$\sum_{i=1}^{d}\gamma_i^j(F_i)dF_i = 0$$

to obtain an upper bound for $r$ with respect to $d$.

More precisely, we can present the *Poincaré-Blaschke method* in the following way: $Z_i(x,y) = \left(\gamma_i^1(F_i(x,y)),\ldots,\gamma_i^r(F_i(x,y))\right)$ define $d$ germs of maps of rank at most 1 from $(\mathbb{C}^2,0)$ to $(\mathbb{C}^r, Z_i(0))$. We consider the non-decreasing sequence of subspaces of $\mathbb{C}^r$ spanned by the osculating spaces of germs of increasing order defined by the $Z_i(x,y)$ for $1\leq i\leq d$, that is

$$\{Z_1(x,y),\ldots,Z_d(x,y)\} = \{Z_i(x,y)\} = \mathbb{C}^{N_0(x,y)}(x,y)$$

where $\dim_{\mathbb{C}}\{Z_i(x,y)\} = N_0(x,y)$ and more briefly

$$\{Z_i(x,y),\partial_x(Z_i),\partial_y(Z_i)\} = \mathbb{C}^{N_1(x,y)}(x,y)\,,$$
$$\{Z_i(x,y),\partial_x(Z_i),\partial_y(Z_i),\partial_x^2(Z_i),\partial_x\partial_y(Z_i),\partial_y^2(Z_i)\} = \mathbb{C}^{N_2(x,y)}(x,y), \text{ etc.}$$

Generically, the maps $(x,y) \longmapsto N_l(x,y)$ are locally constant. Using the equality

$$\sum_{i=1}^{d} Z_i(x,y)dF_i(x,y) = 0$$

and the general position assumption, we can assume that we have

$$\mathbb{C}^{N_0(x,y)}(x,y) = \{Z_3(x,y), \ldots, Z_d(x,y)\}.$$

Indeed, we can assume that $F_1 = x$, $F_2 = y$ and omitting the dependency on $(x,y)$ we get the following system:

$$\begin{cases} Z_1 + Z_3 \partial_x(F_3) + \cdots + Z_d \partial_x(F_d) = 0 \\ Z_2 + Z_3 \partial_y(F_3) + \cdots + Z_d \partial_y(F_d) = 0 \end{cases}$$

Every minor determinant of order 2 of the matrix

$$\begin{pmatrix} 1 & 0 & \partial_x(F_3) & \ldots & \partial_x(F_d) \\ 0 & 1 & \partial_y(F_3) & \ldots & \partial_y(F_d) \end{pmatrix}$$

is invertible in $\mathcal{O}$ and we have $X_i(Z_i) = 0$ for $1 \leq i \leq d$; this enables to check, differentiating several times the previous system, modulo $\{Z_i(x,y)\}$ (resp. modulo $\{Z_i(x,y), \partial_x(Z_i), \partial_y(Z_i)\}$, etc.) that we have

$$\mathbb{C}^{N_1(x,y)}(x,y) = \{Z_3, \ldots, Z_d, \partial_x(Z_4), \ldots, \partial_x(Z_d)\},$$
$$\mathbb{C}^{N_2(x,y)}(x,y) = \{Z_3, \ldots, Z_d, \partial_x(Z_4), \ldots, \partial_x(Z_d), \partial_x^2(Z_5), \ldots, \partial_x^2(Z_d)\}, \text{ etc.}$$

This way we can exhaust the order of osculating spaces and this implies that the preceding sequence is stationary; more precisely we get

$$N_0(x,y) \leq N_1(x,y) \leq \cdots \leq N_{d-3}(x,y) = N_{d-2}(x,y) = \cdots$$

Let $(x_0, y_0)$ be a point in $\mathbb{C}^2$ in the neighborhood of which we assume that the map $(x,y) \longmapsto N_{d-3}(x,y)$ is constant. Using analyticity and the Taylor formula we get $\mathbb{C}^{N_{d-3}(x_0,y_0)}(x_0,y_0) = \mathbb{C}^{N_{d-3}(x,y)}(x,y)$ near $(x_0, y_0)$; moreover, since the family $\left(\gamma_i^j(F_i)\right)_{1 \leq i \leq d, 1 \leq j \leq r}$ is independent, we must have $\mathbb{C}^{N_{d-3}(x,y)}(x,y) = \mathbb{C}^r$ near $(x_0, y_0)$. With the help of the preceding descriptions, we obtain the claimed upper bound

$$r = N_{d-3}(x,y) \leq (d-2) + (d-3) + \cdots + 1 = \frac{(d-1)(d-2)}{2}.$$

If $\text{rk}\,\mathcal{W}(d)$ is *maximal*, using a basis of the space $\mathcal{A}(d)$ of Abelian relations of the $d$-web $\mathcal{W}(d)$ in $(\mathbb{C}^2, 0)$, we can carry on the above construction with $r = \dim_\mathbb{C} \mathcal{A}(d) = \frac{1}{2}(d-1)(d-2)$. Under these hypotheses and using semi-continuity, the previous conditions imply that we have, for $(x,y)$ near $0 \in \mathbb{C}^2$, the following families of subspaces:

$$\{Z_i(x,y)\} = \mathbb{C}^{d-2}(x,y) \subseteq \mathbb{C}^{\frac{1}{2}(d-1)(d-2)}$$
$$\{Z_i(x,y), \partial_x(Z_i), \partial_y(Z_i)\} = \mathbb{C}^{2d-5}(x,y) \subseteq \mathbb{C}^{\frac{1}{2}(d-1)(d-2)}$$

$$\cdots$$

$$\{Z_i(x,y), \partial_x(Z_i), \partial_y(Z_i), \ldots\} = \mathbb{C}^{\frac{1}{2}(d-1)(d-2)}.$$

The two former relations will be used afterwards and the later enables, for instance, to give another proof of proposition 1.

This method, usually presented through successive changes of variables, can be generalized to higher dimensions and gives rise to numerous variants (*cf.* [10,16,18,19]). Some of them will be explored in section 3; furthermore, we will see how to use successfully the geometry of the space of Abelian relations (*i.e.* essentially the projective configuration of the family of points $Z_i(x, y)$) for the algebraization of webs of maximal rank.

*Remark 1.* The Poincaré-Blaschke method, suitably adapted, can be set to work in the $C^\infty$ setting; this shows that theorem 1 holds in this setting.

The method using $\mathcal{D}$-modules goes on that way: let us consider the linear differential system $\mathcal{R}(d)$ whose solutions $(f_1, \ldots, f_d) \in \mathcal{O}^d$ satisfy

$$\mathcal{R}(d) \quad \begin{cases} X_i(f_i) = 0 \text{ for } 1 \leq i \leq d \\ \partial_x(f_1 + \cdots + f_d) = 0 \\ \partial_y(f_1 + \cdots + f_d) = 0; \end{cases}$$

The $d$ former equations are the eikonal solutions of the system $\mathcal{R}(d)$ called system of the *resonance equations* in the $C^\infty$ setting (*cf.* [44]). If we denote by $\mathrm{Sol}\,\mathcal{R}(d)$ the $\mathbb{C}$-vector space of solutions of $\mathcal{R}(d)$, we get an exact sequence of $\mathbb{C}$-vector spaces

$$0 \longrightarrow \mathbb{C}^d \longrightarrow \mathrm{Sol}\,\mathcal{R}(d) \stackrel{\delta}{\longrightarrow} \mathcal{A}(d) \longrightarrow 0$$

where $\delta\big((f_i)_i\big) = \big(\alpha_i'(F_i)\big)_i$ since each $f_i \in \mathcal{O}$ satisfying $X_i(f_i) = 0$ can be expressed as $f_i = \alpha_i(F_i)$ with $\alpha_i \in \mathbb{C}\{z\}$.

With the help of the general position assumption, it can be shown (*cf.* [38] and [41] for details) that the left $\mathcal{D}$-module $\mathcal{R}(d)$ associated to the differential system is *holonomic* and that its characteristic variety has the zero section for support (*cf.* for instance [31] for basic results about $\mathcal{D}$-modules).

Using a recurrence on $d$, which amounts to extracting sub-webs of $\mathcal{W}(d)$, we estimate the *multiplicity* mult $\mathcal{R}(d)$ of $\mathcal{R}(d)$, using the matrices of symbols associated to the differential system $\mathcal{R}(d)$. This gives a control on the rank since the preceding exact sequence yields the equality

$$\mathrm{rk}\,\mathcal{W}(d) = \dim_{\mathbb{C}} \mathcal{A}(d) = \dim_{\mathbb{C}} \mathrm{Sol}\,\mathcal{R}(d) - d = \mathrm{mult}\,\mathcal{R}(d) - d.$$

This method generalizes to higher dimensions (*cf.* [41]) and can be used for singular webs, *via* for instance the Kashiwara index theorem (*cf.* [45]) and more generally the algebro-geometric description of the *characteristic cycle* of $\mathcal{R}(d)$ (*cf.* for instance sect. 3.3).

## 2.2 Algebraic webs

Let $C \subset \mathbb{P}^2$ be a *reduced* algebraic curve of degree $d$, not necessarily irreducible and possibly singular. A generic line $l(0) \in G(1, \mathbb{P}^2) = \check{\mathbb{P}}^2$ intersects $C$ through $d$ distinct smooth points $p_i(0)$ which, *using duality*, give $d$ different lines in $\check{\mathbb{P}}^2$ passing through $l(0)$. This construction can be completed for lines $l(x)$ near $l(0)$:

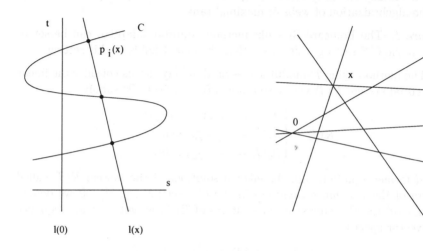

Figure 7.

to obtain a $d$-web $\mathcal{L}_C(d)$ in $(\mathbb{C}^2, 0) = (\check{\mathbb{P}}^2, l(0))$ which is *linear* since its leaves are straight lines, not necessarily parallel to one-another. More precisely, if $x$ is near $0 \in \mathbb{C}^2$ the leaves passing through $x$ correspond, by duality, to the points $p_i(x)$ satisfying

$$l(x) \cap C = \sum_{i=1}^{d} p_i(x) \quad \text{as 0-cycles in } C.$$

We call $\mathcal{L}_C(d)$ *"the" algebraic web associated to* $C \subset \mathbb{P}^2$. If $C$ contains no straight line, the leaves of $\mathcal{L}_C(d)$ are the tangents of the dual curve $\check{C} \subset \check{\mathbb{P}}^2$ of $C \subset \mathbb{P}^2$, otherwise, these belong to the corresponding pencils of rays.

In a suitable coordinate system, we have $d$ local branches $p_i = (F_i, \xi_i(F_i))$ over $\check{\mathbb{P}}^2$ where $f(s, t) = \prod_{i=1}^{d}(t - \xi_i(s)) = 0$ is an affine equation of $C$, $l(x) =$

$l(x_1, x_2) = \{s = x_1 + tx_2\}$ and implicitly $F_i(x) = x_1 + \xi_i(F_i(x)).x_2$.

Let $\omega_C$ be the dualizing sheaf of $C$ (if $C$ is smooth, then $\omega_C = \Omega_C^1$); its global sections, that is Abelian differential 1-forms over $C$, form a $\mathbb{C}$-vector space $H^0(C, \omega_C)$ spanned by

$$r(s,t) \frac{ds}{\partial_t(f)} \text{ where } r \in \mathbb{C}[s,t] \text{ and } \deg r \leq d - 3.$$

We have a $\mathbb{C}$-linear map

$$A : H^0(C, \omega_C) \longrightarrow A(\mathcal{L}_C(d))$$

defined by $A(\omega) = (g_i(F_i))$ where $\omega = g_i(s)ds$ in the neighborhood of $p_i(0)$. Effectively, Abel's theorem (which justifies the terminology), gives (cf. [1] and more generally [32])

$$\text{Trace}(\omega) := \sum_{i=1}^d p_i^*(\omega) = \sum_{i=1}^d g_i(F_i)dF_i = 0.$$

This can be checked through direct computation, for fixed $(x, y)$ in $\check{\mathbb{P}}^2$, using the residue formula in $\mathbb{P}^1$. Moreover, it can be shown that $A$ is an isomorphism (cf. [40]).

The description of $H^0(C, \omega_C)$ shows that the inequality of theorem 1 is optimal, and we get the following result:

**Proposition 2** *The rank of the algebraic web $\mathcal{L}_C(d)$ associated to the curve $C \subset \mathbb{P}^2$ is maximal and equal to the arithmetic genus of $C$, in other words*

$$\text{rk}\, \mathcal{L}_C(d) = \dim_{\mathbb{C}} H^0(C, \omega_C) = \frac{(d-1)(d-2)}{2}.$$

The preceding properties raise two fundamental problems:

1. Determine the webs in $(\mathbb{C}^2, 0)$ which are *linearizable* (*i.e.* which can be made linear through a local analytical isomorphism of $\mathbb{C}^2$).

2. Determine the webs $\mathcal{W}(d)$ in $(\mathbb{C}^2, 0)$ of maximal rank and which are *algebraizable* (*i.e.* such that $\mathcal{W}(d) = \mathcal{L}_C(d)$ up to a local isomorphism).

Let $\mathcal{W}(d)$ be a $d$-web in $(\mathbb{C}^2, 0)$, we say that an Abelian relation $\sum_{i=1}^d g_i(F_i)dF_i = 0$ of $\mathcal{W}(d)$ is *algebraizing* if $g_i \neq 0$ for $1 \leq i \leq d$.

It can be checked that a $d$-web $\mathcal{W}(d)$ in $(\mathbb{C}^2, 0)$ of maximal rank has at least one algebraizing Abelian relation provided $d \geq 3$.

Problem 2 has been (partly) solved through the next theorem, which is in fact a reciprocal of Abel's theorem (cf. [48] and [21] for $d = 4$, and [32] concerning the general situation which will be studied in more detail in section 3):

**Theorem 2.2 (Lie, Darboux, Griffiths)** *Let $\mathcal{L}(d)$ be a linear $d$-web in $(\mathbb{C}^2, 0)$ with $d \geq 3$ with at least one algebraizing Abelian relation. Then $\mathcal{L}(d)$ is algebraic. In other words, there exists a reduced algebraic curve $C \subset \mathbb{P}^2$ with degree $d$, not necessarily irreducible and possibly with singularities, such that $\mathcal{L}(d) = \mathcal{L}_C(d)$.*

*In particular, every $d$-web in $(\mathbb{C}^2, 0)$ of maximal rank which can be linearized is algebraizable.*

Let $\mathcal{W}(d)$ be a $d$-web in $(\mathbb{C}^2, 0)$. For $d = 1$ (resp. 2), we already mentioned that $\mathcal{W}(d)$ can always be linearized; the same holds if $d = 3$ provided that $\mathrm{rk}\,\mathcal{W}(3) = 1$ (cf. proposition 1).

Let $\mathcal{W}(4)$ be a 4-web in $(\mathbb{C}^2, 0)$ with *maximal* rank, which means $\mathrm{rk}\,\mathcal{W}(4) = 3$. Using a basis of the space $\mathcal{A}(4)$ of Abelian relations of $\mathcal{W}(4)$ and the notations and observations preceding remark 1, we can construct a map

$$\ell : (\mathbb{C}^2, 0) \longrightarrow (\check{\mathbb{P}}^2, \ell(0)) = (\mathbb{C}^2, 0)$$

where $\ell(x, y) = \{Z_i(x, y)\} = \mathbb{P}^1(x)$ and $\{Z_i(x, y), \partial_x(Z_i), \partial_y(Z_i)\} = \mathbb{P}^2 = \mathbb{P}(\mathcal{A}(4))$. Therefore $\ell$ is an isomorphism. Moreover, due to its definition, the "image" web $\ell_*(\mathcal{W}(4))$ is *linear* and of maximal rank, and therefore it is algebraic. This fundamental construction is due to H. Poincaré (cf. [57]); it will be generalized in the next chapter and gives for the moment the following result:

**Proposition 3** *Every 4-web in $(\mathbb{C}^2, 0)$ of maximal rank 3 can be linearized.*

There are webs of maximal rank which cannot be algebraized. The only known example dates back to 1936; it is Bol's 5-web $\mathcal{B}$ (cf. [11]). Its rank is 6 and it cannot be linearized. We will come back to it later (cf. (sect. 3.3 and 3.4) and (sect. 3.1, 3.2 and 3.3)), but for the moment, here is a description. This web is formed by 4 pencils of lines in $\mathbb{C}^2$ whose vertices are in general position, and for any generic $z$, the fifth leaf through $z$ is the *only* conic passing through the four vertices and $z$ (cf. figure 8).

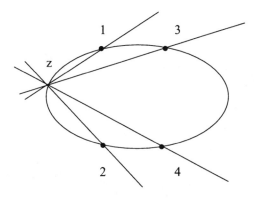

Figure 8.

## 2.3 About 3-webs in $(\mathbb{C}^2, 0)$; hexagonal webs

We have seen (*cf.* sect. 1) that the case of 3-webs in $(\mathbb{C}^2, 0)$ is a particular one since the rank of such a web is either 0 or 1. Another characterization of this dichotomy is due to G. Thomsen; historically, it is the first one and dated 1927 (*cf.* [61] and also [38]).

Let $\mathcal{W}$ be a 3-web in $(\mathbb{C}^2, 0)$. For $z$ near $0 \in \mathbb{C}^2$ and traveling along the leaves of $\mathcal{W}$ around $z$, we build the construction suggested by the following drawing:

We say that the web $\mathcal{W}$ is *hexagonal* (or satisfies Thomsen's property) if, for every $z$ near $0 \in \mathbb{C}^2$ and every $(A, i)$ near $z$ on one of the leaves passing through $z$, the points $G(z; (A, i))$ and $(A, i)$ are the same. In other words, $\mathcal{W}$ is hexagonal if for all $z$ near $0 \in \mathbb{C}^2$, every "hexagon" built around $z$ is closed.

For instance, the 3-web $\mathcal{H}$ defined by $x = cst$, $y = cst$ and $x + y = cst$ is hexagonal:

Here is another construction due to W. Blaschke which gives a third characterization of this dichotomy, using the Pfaff forms $\omega_i$ defining the 3-web $\mathcal{W}$ (*cf.* [10,8,9]).

We assume that the forms $\omega_i$ have been normalized so that we have $\omega_1 + \omega_2 + \omega_3 = 0$ since $\omega$ and $\rho\omega$ define the same foliation if $\rho \in \mathcal{O}^\star$. The 2-form

$$\Omega := \omega_1 \wedge \omega_2 = \omega_2 \wedge \omega_3 = \omega_3 \wedge \omega_1$$

is non-singular, hence $d\omega_i = h_i.\Omega$ for $1 \leq i \leq 3$. Further, there exists a Pfaff

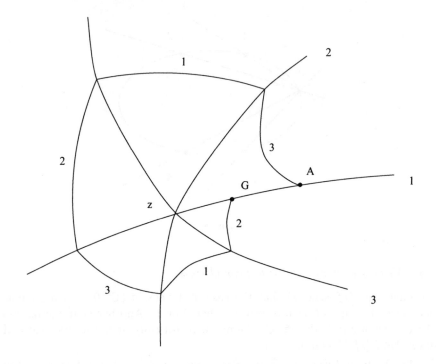

Figure 9.

form $\gamma$ such that $d\omega_i = \gamma \wedge \omega_i$ for $1 \leq i \leq 3$, that is

$$\gamma := h_2\omega_1 - h_1\omega_2 = h_3\omega_2 - h_2\omega_3 = h_1\omega_3 - h_3\omega_1.$$

The 2-form $K(\mathcal{W}) := d\gamma$ is said to be the (Blaschke) *curvature* of $\mathcal{W}$; indeed, it can be checked that $d\gamma$ depends only on $\mathcal{W}$ and not on the *normalized* choice of the $\omega_i$.

If $\mathcal{H}$ is defined by $dx$, $dy$ and $-d(x+y)$ then $\mathcal{H}$ has *zero curvature* since $h_1 = h_2 = h_3 = 0$.

If $\mathcal{W}_F$ is defined by $x = cst$, $y = cst$ and $F(x,y) = cst$, we have the normalization $\partial_x(F)dx + \partial_y(F)dy - dF = 0$ and we can check that

$$K(\mathcal{W}_F) = \partial_x \partial_y \left( \log \frac{\partial_x(F)}{\partial_y(F)} \right) dx \wedge dy.$$

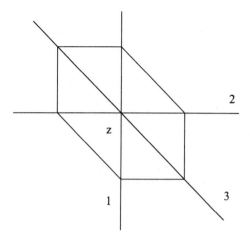

Figure 10.

With the help of what has been said so far, we obtain the following theorem which sums up and enlightens properties presented previously:

**Theorem 2.3** *Concerning a 3-web $\mathcal{W}$ in $(\mathbb{C}^2, 0)$, the following assertions are equivalent:*

*(i)* rk $\mathcal{W} = 1$;

*(ii)* $\mathcal{W}$ *is parallelizable (i.e. defined up to a local analytical isomorphism of $(\mathbb{C}^2, 0)$, by $x = cst$, $y = cst$ and $x + y = cst$);*

*(iii)* $\mathcal{W}$ *is hexagonal;*

*(iv)* $\mathcal{W}$ *has zero curvature.*

Here are some examples and remarks.

Outside its singular locus, the 3-web $\mathcal{P}(3)$ in $\mathbb{C}^2$ defined by $x$, $y$ and $y/x$ is hexagonal since $x.(1/y).(y/x) = 1$ and then $\log x - \log y + \log(y/x) = 0$ proving *(i)* or because $\partial_x \partial_y \left( \log \dfrac{-y/x^2}{1/x} \right) = 0$ proving *(iv)*. In fact, this web is the 3-web in $\check{\mathbb{P}}^2$ formed by the 3 pencils of lines in $\mathbb{P}^2$ with vertices $[0, 0, 1]$, $[0, 1, 0]$ and $[1, 0, 0]$.

More generally, 3 distinct points in $\check{\mathbb{P}}^2$ *generically* define a hexagonal linear 3-web $\mathcal{P}(3)$; it can be checked through Thomsen's characterization or through the fact that $\mathcal{P}(3) = \mathcal{L}_C(3)$ where $C \subset \mathbb{P}^2$ is the union of the 3 lines corresponding, by duality, to the 3 distinct points in $\check{\mathbb{P}}^2$.

It can be shown that near $0 \in \mathbb{C}^2$, the 3-web defined by $x = cst$, $y = cst$ and the bundle of circles passing through $(0,1)$ and $(1,0)$ is *not* hexagonal; this can be seen through (*iii*) and a precise drawing, or through (*iv*) and a computation of the curvature.

If $C$ is a smooth cubic in $\mathbb{P}^2$, we know that $C$ can be equipped with a group structure. Associativity for this law and the hexagonal property for the 3-web $\mathcal{L}_C(3)$ in $\check{\mathbb{P}}^2$ can be put together.

It is tempting to study properties of $d$-webs in $(\mathbb{C}^2, 0)$ through their sub-webs. Concerning this point of view, it is interesting to restate the following result of G. Bol (*cf.* [10,20]):

**Theorem 2.4** *Let $\mathcal{H}(d)$ be a hexagonal $d$-web in $(\mathbb{C}^2, 0)$ with $d \geq 3$ (i.e. every extracted 3-web of $\mathcal{H}(d)$ is hexagonal). Then, up to a local analytical isomorphism of $(\mathbb{C}^2, 0)$, only two possibilities occur:*
$\mathcal{H}(d) = \mathcal{P}(d)$ *where $\mathcal{P}(d)$ is generated by $d$ distinct points in $\check{\mathbb{P}}^2$,*
*or*
$\mathcal{H}(5) = \mathcal{B}$ *where $\mathcal{B}$ is Bol's 5-web.*

As a continuation, one should refer to work made by I. Nakai (*cf.* [54]) and V.V. Goldberg (*cf.* [29,30]) concerning geometrical properties of 4-webs in $\mathbb{R}^2$ whose extracted 3-webs share the same curvature $K$.

If $\mathcal{W}_{(F_1, F_2, F_3)}$ is a 3-web in $(\mathbb{C}^2, 0)$ defined by the equations $F_i(x, y) = cst$, it can be useful to state, as an ending for this paragraph, that G. Mignard established in [52] (cf. also [53]) an explicit formula for the curvature $K(\mathcal{W}_{(F_1, F_2, F_3)})$. This formula, however complicated, can be expressed using only partial derivatives of the $F_i$ for $1 \leq i \leq 3$.

## 2.4  Linearizing webs in $\mathbb{C}^2$

The goal here is to give conditions for a $d$-web $\mathcal{W}(d)$ in $(\mathbb{C}^2, 0)$ implying that it is linearizable, which means that, up to a local analytical isomorphism of $(\mathbb{C}^2, 0)$, $\mathcal{W}(d)$ is linear (*i.e.* its leaves are straight lines, not necessarily parallel to one-another). A description of the eventual linearizations is also envisioned.

The following results hold (*cf.* [39] for details):

**Lemma 1** *Up to an automorphism of $\check{\mathbb{P}}^2$, local analytical isomorphisms of the form $\phi : (\mathbb{C}^2, 0) \longrightarrow (\check{\mathbb{P}}^2, 0)$ correspond to analytical solutions $(P_0, P_1, P_2, P_3) \in \mathcal{O}^4$ of the following quasi-linear differential system:*

$$\begin{cases} \partial_x^2(P_2) - 2\partial_x\partial_y(P_1) + 3\partial_y^2(P_0) + 6P_0\partial_x(P_3) - 3P_2\partial_y(P_0) \\ \quad -3P_0\partial_y(P_2) + 3P_3\partial_x(P_0) + 2P_1\partial_y(P_1) - P_1\partial_x(P_2) = 0 \\ \\ 3\partial_x^2(P_3) - 2\partial_x\partial_y(P_2) + \partial_y^2(P_1) - 6P_3\partial_y(P_0) + 3P_1\partial_x(P_3) \\ \quad +3P_3\partial_x(P_1) - 3P_0\partial_y(P_3) - 2P_2\partial_x(P_2) + P_2\partial_y(P_1) = 0 \,. \end{cases} \quad (\mathcal{S})$$

*To be more precise, each $\phi$ can be obtained through a linear differential system based on the $(P_0, P_1, P_2, P_3) \in \mathcal{O}^4$ and $(\mathcal{S})$ describes the integrability conditions for this system.*

Let $\mathcal{W}(d)$ be a $d$-web in $(\mathbb{C}^2, 0)$; after an eventual *linear* coordinate change, we can assume that the vector fields associated to this web are of the form

$$X_i = \partial_x + b_i\partial_y \text{ for } 1 \leq i \leq d;$$

the general position assumption then becomes $b_i(0) \neq b_j(0)$ for $1 \leq i < j \leq d$.

We can prove that $\mathcal{W}(d)$ is *linear* if and only if $X_i(b_i) = 0$ for $1 \leq i \leq d$ (*i.e.* the $b_i$ satisfy the Burgers-Hopf equation $\dfrac{\partial b}{\partial x} + b\dfrac{\partial b}{\partial y} = 0$).

**Theorem 2.5** *A $d$-web $\mathcal{W}(d)$ in $(\mathbb{C}^2, 0)$ can be linearized if and only if the differential system*

$$\begin{cases} \partial_x^2(P_2) - 2\partial_x\partial_y(P_1) + 3\partial_y^2(P_0) + 6P_0\partial_x(P_3) - 3P_2\partial_y(P_0) \\ \quad -3P_0\partial_y(P_2) + 3P_3\partial_x(P_0) + 2P_1\partial_y(P_1) - P_1\partial_x(P_2) = 0 \\ \\ 3\partial_x^2(P_3) - 2\partial_x\partial_y(P_2) + \partial_y^2(P_1) - 6P_3\partial_y(P_0) + 3P_1\partial_x(P_3) \\ \quad +3P_3\partial_x(P_1) - 3P_0\partial_y(P_3) - 2P_2\partial_x(P_2) + P_2\partial_y(P_1) = 0 \\ \\ P_0 + P_1 b_i + P_2 b_i^2 + P_3 b_i^3 = X_i(b_i) \text{ for } 1 \leq i \leq d \end{cases} \quad (\star)_d$$

*admits an analytical solution $(P_0, P_1, P_2, P_3) \in \mathcal{O}^4$. In particular, if $d \geq 4$, a web $\mathcal{W}(d)$ admits at most one linearization up to a projective transformation.*

*The Gronwall conjecture* also called the fundamental "theorem" of nomography (result announced in 1912 at the end of the introduction of [35]) can be phrased, using the preceding notations, in the following form: *Let $\mathcal{W}(3)$ be a 3-web in $(\mathbb{C}^2, 0)$ with rank zero, then the system $(\star)_3$ admits at most one analytical solution.*

For a different point of view on the linearization of 3-webs in $(\mathbb{C}^2, 0)$, one can refer to [28].

Starting with a $d$-web $\mathcal{W}(d)$ in $(\mathbb{C}^2, 0)$, we can construct a *unique* polynomial

$$P_{\mathcal{W}(d)} = P_0 + P_1 b + \cdots + P_{d-1} b^{d-1} \in \mathcal{O}[b]$$

such that

$$P_{\mathcal{W}(d)}(x, y; b_i) = X_i(b_i) \text{ for } 1 \leq i \leq d.$$

Moreover, we can prove that the $d$ leaves of $\mathcal{W}(d)$ are graphs of elements $y_i \in \mathbb{C}\{x\}$ satisfying the differential equation of the second order

$$y'' = P_{\mathcal{W}(d)}(x, y; y').$$

What precedes can be summarized through the following result:

**Corollary 1** *If $d \geq 4$, a $d$-web $\mathcal{W}(d)$ in $(\mathbb{C}^2, 0)$ can be linearized if and only if $\deg P_{\mathcal{W}(d)} \leq 3$ and $(P_0, P_1, P_2, P_3)$ satisfy the differential system $(\mathcal{S})$.*

This enables to find again classical results of R. Liouville, A. Tresse, E. Cartan et V. Arnold about the linearization of the second order differential equation $y'' = P_{\mathcal{W}(d)}(x, y; y')$ (*cf.* respectively [49,62,12] and [5]). Further, using the preceding lemma, the differential system $(\mathcal{S})$ can be used to describe all possible linearizations.

This corollary ensures for instance that Bol's 5-web $\mathcal{B}$ in $(\mathbb{C}^2, 0)$ (*cf.* sect. 2.2 and 2.3) cannot be linearized. Indeed, it can be proven that $\deg P_{\mathcal{B}} = 4$.

More generally speaking, the polynomial $P_{\mathcal{W}(d)}$ associated with a $d$-web $\mathcal{W}(d)$ in $(\mathbb{C}^2, 0)$ gathers differential invariants of $\mathcal{W}(d)$ worth studying. In this perspective and in the light of general results of A. Tresse concerning the invariants of a second-order differential equation (*cf.* [63]), the examination of the particular case $y'' = P_{\mathcal{W}(d)}(x, y; y')$ would be welcome.

## 3  Webs of codimension $n$ in $\mathbb{C}^{kn}$

### 3.1  *General properties of webs $\mathcal{W}(d, k, n)$ of codimension $n$ in $\mathbb{C}^{kn}$*

A $d$-web $\mathcal{W}(d, k, n)$ in $(\mathbb{C}^{kn}, 0)$ is formed by $d$ complex analytic foliations of codimension $n$ in $(\mathbb{C}^{kn}, 0)$ in general position. More precisely, if $\mathcal{O} := \mathbb{C}\{x_1, \ldots, x_{kn}\}$ denotes the ring of convergent power series in $kn$ variables, then the web $\mathcal{W}(d,k,n)$ is defined by $d$ *families* of leaves of codimension $n$ in general position; these are germs of level sets defined by

$$\begin{cases} F_{i_1}(x) = cst \\ \quad \cdots \\ F_{i_n}(x) = cst \end{cases}$$

where $x = (x_1, \ldots, x_{kn})$ and $F_{i_m} \in \mathcal{O}$ satisfy $F_{i_m}(0) = 0$.

For $1 \leq i \leq d$ and $x$ near $0 \in \mathbb{C}^{kn}$, the "*normals*" of the leaves passing through $x$

$$\Omega_i(x) = \bigwedge_{m=1}^{n} dF_{i_m}(x)$$

define $d$ points of the Grassmann manifold $G(n-1, \mathbb{P}^{kn-1})$ of subspaces of dimension $(n-1)$ in $\mathbb{P}^{kn-1}$ which can be considered, *via* Plücker's embedding, as points in $\mathbb{P}^{\binom{kn}{n}-1}$; these normals depend only on the web $\mathcal{W}(d,k,n)$ and not on the choice of the $F_{i_m}$, and the *general position* amounts to

$$\Omega_{i(1)}(0) \wedge \cdots \wedge \Omega_{i(j)}(0) \neq 0 \text{ for all } 1 \leq i(1) < \cdots < i(j) \leq d \text{ with } j \leq k.$$

If $d = 1$ (resp. 2, ..., resp. $k$), the local inverse mapping theorem implies that the local model for a web $\mathcal{W}(d,k,n)$ is given by the following families of $(k-1)n$-planes in $\mathbb{C}^{kn}$:

$$\{x_1 = cst, \ldots, x_n = cst\},$$

$$(\text{resp. } \{x_1 = cst, \ldots, x_n = cst\} \text{ and } \{x_{n+1} = cst, \ldots, x_{2n} = cst\},$$

$$\cdots$$

$$\text{resp. } \{x_1 = cst, \ldots, x_n = cst\}, \ldots, \{x_{(k-1)n+1} = cst, \ldots, x_{kn} = cst\});$$

which implies that the study of possible configurations for webs $\mathcal{W}(d,k,n)$ is interesting only for $d \geq k+1$.

The main invariant of a web $\mathcal{W}(d,k,n)$ in $(\mathbb{C}^{kn}, 0)$ is related with the notion of Abelian relation already developed for webs in $(\mathbb{C}^2, 0)$ (*cf.* sect. 2.1)). A $d$-uple $(g_1(F_{i_1}, \ldots, F_{i_n}), \ldots, g_d(F_{i_1}, \ldots, F_{i_n})) \in \mathcal{O}^d$ with $g_i \in \mathbb{C}\{z\} = \mathbb{C}\{z_1, \ldots, z_n\}$ satisfying

$$\sum_{i=1}^{d} g_i(F_{i_1}, \ldots, F_{i_n}) dF_{i_1} \wedge \cdots \wedge dF_{i_n} = 0$$

is called an *Abelian relation of degree $n$* for the web $\mathcal{W}(d,k,n)$; the $\mathbb{C}$-vector space of Abelian relations of degree $n$ of the web $\mathcal{W}(d,k,n)$ will be denoted by

$$\mathcal{A}^n = \big\{ \big( g_i(F_{i_1}, \ldots, F_{i_n}) \big) \mid g_i \in \mathbb{C}\{z\} \text{ and } \sum_{i=1}^{d} g_i(F_{i_1}, \ldots, F_{i_n}) \, \Omega_i = 0 \big\}.$$

Let $\big( \gamma_i^j (F_{i_1}, \ldots, F_{i_n}) \big)_{1 \le i \le d, 1 \le j \le r_n}$ be an independent family in $\mathcal{A}^n$. The formulas $Z_i(x) = \big( \gamma_i^1 (F_{i_1}, \ldots, F_{i_n}), \ldots, \gamma_i^{r_n} (F_{i_1}, \ldots, F_{i_n}) \big)$ give $d$ germs of maps with rank at most $n$

$$Z_i : (\mathbb{C}^{kn}, 0) \longrightarrow (\mathbb{C}^{r_n}, Z_i(0)).$$

Using the *vector identity*

$$\sum_{i=1}^{d} Z_i(x).\Omega_i(x) = 0$$

defined for $x$ near $0 \in \mathbb{C}^{2n}$ and generalizing the Poincaré-Blaschke method (*cf.* sect. 2.1), we can prove the following *Chern and Griffiths upper bound* (*cf.* [16] in case $n = 1$ and [19] for the general case):

**Theorem 3.1** *Let* $\mathcal{W}(d, k, n)$ *be a d-web of codimension n in* $(\mathbb{C}^{kn}, 0)$, *we have the following optimal upper bound:*

$$\mathrm{rk}_n \, \mathcal{W}(d, k, n) := \dim_{\mathbb{C}} \mathcal{A}^n \le \pi(d, k, n)$$

*where* $\pi(d, k, n)$ *is the generalized Castelnuovo number defined through*

$$\pi(d, k, n) = \{d - kn + n - 1\} + \binom{n}{1} \{d - k(n + 1) + n\}$$
$$+ \binom{n + 1}{2} \{d - k(n + 2) + n + 1\} + \cdots$$

*with the convention that summation concerns only the positive terms.*

It can be checked that the integer $\mathrm{rk}_n \, \mathcal{W}(d, k, n)$ defined above is an analytic invariant of the web $\mathcal{W}(d, k, n)$ which does not depend on the choice of the $F_{i_m}$. It is called the *n-rank* of the web.

If $n = 1$, $\pi(d, k, 1) = \{d - k\} + \{d - 2k + 1\} + \{d - 3k + 2\} + \cdots$ is the standard Castelnuovo number; we will come back to it later.

In other respects, for $k = 2$ we have

$$\pi(d, 2, n) = \binom{d - 1}{n + 1} := \frac{(d - 1)(d - 2) \cdots (d - n - 1)}{(n + 1)!}.$$

In particular, this proves again that $\pi(d, 2, 1) = (d - 1)(d - 2)/2$ is an upper bound for the 1-rank, that is the rank of $d$-webs in $(\mathbb{C}^2, 0)$.

The above-mentioned upper bound can also be obtained, in case $n = 1$, through basic properties of $\mathcal{D}$-modules using a recurrence on $d$ (cf. [41]); this is in fact a generalization of the method sketched at the end of sect. 2.1 to prove the upper bound of the rank of webs in $(\mathbb{C}^2, 0)$.

*Remark 2.* The upper bound of Chern and Griffiths presented above probably holds in the $C^\infty$ setting.

In a general way, and carrying on an idea of P.A. Griffiths (cf. [33]), we can associate to any $d$-web $\mathcal{W}(d, k, n)$ in $(\mathbb{C}^{kn}, 0)$ Abelian relations of degree $p$ for $1 \leq p \leq n$. This way we get $\mathbb{C}$-vector spaces $\mathcal{A}^p$ of Abelian relations of degree $p$ of the web $\mathcal{W}(d, k, n)$. We have to point out that the exterior differential induces a *complex* $(\mathcal{A}^\bullet, \delta)$ and that, for $1 \leq p \leq n$, a variant of the Poincaré-Blaschke method can be established to obtain an *optimal bound* for the $p$-rank $r_p$ of $\mathcal{W}(d, k, n)$ with the definition $r_p := \dim_\mathbb{C} \mathcal{A}^p$. These ranks are analytical invariants of the web $\mathcal{W}(d, k, n)$ and their bounds $\pi_p(d, k, n)$ are the *generalized Castelnuovo numbers*. For $p = n$, we recover the previous ones.

We will come back to these objects in more detail at the end of this chapter, concerning the special case of $d$-webs $\mathcal{W}(d, 2, n)$ in $(\mathbb{C}^{2n}, 0)$.

As we will see soon, projective algebraic geometry gives models of webs $\mathcal{W}(d, k, n)$ of codimension $n$ in $(\mathbb{C}^{kn}, 0)$. This is one reason to particularize the study of webs such that the codimension of leaves divides the dimension of the ambient space.

Effectively, every *"general enough"* projective algebraic variety $V_n$ in $\mathbb{P}^{n+k-1}$ of dimension $n$ and degree $d$ generically determine some $(k-1)$-planes in the Grassmann manifold $G(k-1, \mathbb{P}^{n+k-1})$ which define locally in $\mathbb{C}^{kn}$ a $d$-web of codimension $n$.

To be more precise, if $V_n$ is a reduced algebraic variety in $\mathbb{P}^{n+k-1}$ of *pure* dimension $n$ which is *non degenerate* (*i.e.* not contained in an hyperplane of $\mathbb{P}^{n+k-1}$), not necessarily irreducible and possibly singular and of degree $d$.

A generic $(k-1)$-plane $\mathbb{P}^{k-1}(0) \in G(k-1, \mathbb{P}^{n+k-1})$ intersects $V_n$ transversally through $d$ distinct smooth points $p_i(0)$ and we assume *from now on* that these $d$ points are in *general position* in $\mathbb{P}^{k-1}(0)$. This is always the case for $k = 2$. If $k \geq 3$ such a generic $(k-1)$-plane exists for instance if we assume $V_n$ to be irreducible, and especially if $V_n$ is smooth and connected.

If we identify locally $(G(k-1, \mathbb{P}^{n+k-1}), \mathbb{P}^{k-1}(0)) = (\mathbb{C}^{kn}, 0)$, the Schubert variety $\sigma_{p_i(0)}$ of $(k-1)$-planes in $\mathbb{P}^{n+k-1}$ passing through $p_i(0)$ corresponds to a $(k-1)n$-plane in $\mathbb{C}^{kn}$ passing through 0. This construction can be carried out for points $x$ near $0 \in \mathbb{C}^{kn}$ and with the help of the general position assumption we get a *linear d-web* $\mathcal{L}_{V_n}(d, k, n)$ in $(\mathbb{C}^{kn}, 0)$ whose leaves passing through $x$ are the $\sigma_{p_i(x)}$ and which we will call *"the" algebraic web associated with* $V_n \subset \mathbb{P}^{n+k-1}$.

In a suitable coordinate system, we have $d$ local branches

$$p_i = (F_{i_1}, \ldots, F_{i_n}, \xi_{i_1}(F_{i_1}, \ldots, F_{i_n}), \ldots, \xi_{i_{k-1}}(F_{i_1}, \ldots, F_{i_n}))$$

over $G(k-1, \mathbb{P}^{n+k-1})$ where

$$\mathbb{P}^{k-1}(x) \cap V_n = \sum_{i=1}^{d} p_i(x)$$

as 0-cycles in $V_n$.

Let $\omega_{V_n}^n$ stand for the *Barlet sheaf* of $V_n$ in maximal degree $n$ (*cf.* [6]); this sheaf is $\mathcal{O}_{V_n}$-coherent and as a consequence its global sections $H^0(V_n, \omega_{V_n}^n)$, in other words the Abelian differential $n$-forms on $V_n$, form a finite dimensional $\mathbb{C}$-vector space. If $V_n$ is smooth, we recall that $\omega_{V_n}^n = \Omega_{V_n}^n$.

We can define a $\mathbb{C}$-linear map

$$A^n : H^0(V_n, \omega_{V_n}^n) \longrightarrow A^n(\mathcal{L}_{V_n}(d, k, n))$$

through $A(\omega) = \big(g_i(F_{i_1}, \ldots, F_{i_n})\big)$ where $\omega = g_i(s_1, \ldots, s_n)ds_1 \wedge \cdots \wedge ds_n$ near $p_i(0)$. Indeed,

$$\text{Trace}(\omega) := \sum_{i=1}^{d} p_i^*(\omega) = \sum_{i=1}^{d} g_i(F_{i_1}, \ldots, F_{i_n})dF_{i_1} \wedge \cdots \wedge dF_{i_n} = 0$$

since the properties of $\omega_{V_n}^n$ imply that the $n$-form $\text{Trace}(\omega)$ admits a holomorphic extension to $G(k-1, \mathbb{P}^{n+k-1})$. This constitutes a generalization of Abel's theorem.

Moreover, it can be proved that $A^n$ is an *isomorphism*. It is enough to pass through the hypersurface case, using a generic projection to some $\mathbb{P}^{n+1}$, in which case the method of [40] can be adapted (case $k = 2$, $n = 1$). Indeed, we can check that if $V_n = \{f(s, t) = f(s_1, \ldots, s_n, t) = 0\}$ with $\deg f = d$, then $H^0(V_n, \omega_{V_n}^n)$ is spanned over $\mathbb{C}$ by

$$r(s, t)\frac{ds_1 \wedge \cdots \wedge ds_n}{\partial_t(f)} \text{ where } r \in \mathbb{C}[s, t] \text{ and } \deg r \le d - n - 2.$$

As a consequence, we deduce from what precedes and the Chern and Griffiths inequality the following result:

**Theorem 3.2** *Under the preceding assumptions, if $\mathcal{L}_{V_n}(d,k,n)$ is the algebraic web associated with $V_n \subset \mathbb{P}^{n+k-1}$, we have*

$$\mathrm{rk}_n\, \mathcal{L}_{V_n}(d,k,n) = \dim_{\mathbb{C}} H^0(V_n, \omega_{V_n}^n).$$

*In particular, this gives the following optimal upper bound:*

$$\dim_{\mathbb{C}} H^0(V_n, \omega_{V_n}^n) \leq \pi(d,k,n).$$

*Moreover, if $k = 2$ (i.e. if $V_n$ is a reduced algebraic hypersurface in $\mathbb{P}^{n+1}$ of degree $d$), then we have*

$$\mathrm{rk}_n\, \mathcal{L}_{V_n}(d,k,n) = \pi(d,2,n) = \binom{d-1}{n+1}.$$

This theorem gives a new proof for Castelnuovo's inequality published in 1889 (*cf.* [13] and [34]):

*If $V_1 \subset \mathbb{P}^k$ is a smooth connected non degenerate curve of degree $d$ and genus $g(V_1) = \dim_{\mathbb{C}} H^0(V_1, \Omega_{V_1}^1)$, then the following optimal inequality holds:*

$$g(V_1) \leq \pi(d,k,1) = \{d-k\} + \{d - 2k + 1\} + \{d - 3k + 2\} + \cdots.$$

The bound of theorem 7 appears also in a work done by J. Harris (*cf.* [37]) as a generalization of G. Castelnuovo's method.

All of the previous upper bounds are *optimal*. Indeed, we can always construct algebraic varieties $V_n \subset \mathbb{P}^{n+k-1}$ satisfying $\dim_{\mathbb{C}} H^0(V_n, \omega_{V_n}^n) = \pi(d,k,n)$.

Further, some such $V_n \subset \mathbb{P}^{n+k-1}$ are *irreducible*, and even smooth connected ones can be produced; irreducible ones are called *Castelnuovo varieties* (or *extremal varieties*) and possess *remarkable* geometrical properties (*cf.* for instance [34,18] and [37]).

As for webs in $(\mathbb{C}^2, 0)$, the problem of the algebraization of webs $\mathcal{W}(d,k,n)$ of codimension $n$ in $(\mathbb{C}^{kn}, 0)$ is natural, notably concerning those of maximal $n$-rank.

Before carrying on, let us develop a particular example. With the notations preceding theorem 6, proposition 3 due to H. Poincaré generalizes in the following way: if $\mathcal{W}((k-1)n + k + 1, k, n) = \mathcal{W}$ has maximal $n$-rank $\{k\} + \binom{n}{1} = n + k$, then we can construct an isomorphism

$$\phi : (\mathbb{C}^{kn}, 0) \longrightarrow (G(k-1, \mathbb{P}^{n+k-1}), \phi(0)) = (\mathbb{C}^{kn}, 0)$$

by setting $\phi(x) = \{Z_i(x)\} = \mathbb{P}^{k-1}(x)$ since in this case $\mathbb{P}(A^n) = \mathbb{P}^{n+k-1}$.

The "image" web $\phi_*(\mathcal{W})$ is clearly *linear*, of codimension $n$ and with maximal $n$-rank. Further, it is obtained through the knowledge of $d$ germs of analytic smooth sets of dimension $n$ in $\mathbb{P}^{n+k-1}$ which are transverse to $\phi(0)$ at $d$ different points.

A general result of P.A. Griffiths (*cf.* [32]) which we already presented in the special case of webs in $(\mathbb{C}^2, 0)$ (*cf.* theorem 2) implies that the web $\phi_*(\mathcal{W})$ is algebraic. More precisely, all $\{Z_i(x)\}$ are contained in a reduced algebraic variety $V_n \subset \mathbb{P}^{n+k-1}$ of degree $d$ and we have $\phi_*(\mathcal{W}) = \mathcal{L}_{V_n}((k-1)n + k + 1, k, n)$.

In other words, the germ of the map $\phi$ defined above using the space of Abelian relations of degree $n$ of the web $\mathcal{W}$ gives an algebraization of the latter. This proves the following result *à la Poincaré*(!):

**Proposition 4** *Every web* $\mathcal{W}((k-1)n + k + 1, k, n)$ *of maximal $n$-rank can be algebraized.*

Two kinds of webs $\mathcal{W}(d, k, n)$ of codimension $n$ in $(\mathbb{C}^{kn}, 0)$ raise particular interest. On one hand, webs $\mathcal{W}(d, k, 1)$ of codimension 1 in $(\mathbb{C}^k, 0)$ whose algebraic models correspond, via *duality*, with curves in $\mathbb{P}^k$. On the other hand, webs $\mathcal{W}(d, 2, n)$ of codimension $n$ in $(\mathbb{C}^{2n}, 0)$ whose algebraic models come from hypersurfaces of $\mathbb{P}^{n+1}$. These two kinds of webs are generalizations of webs $\mathcal{W}(d, 2, 1)$, that is the $d$-webs in $(\mathbb{C}^2, 0)$ we studied before.

The remaining paragraphs of this chapter is devoted to the study of webs $\mathcal{W}(d, 2, n)$.

In the same way as for webs in $(\mathbb{C}^2, 0)$, two main problems arise concerning $\mathcal{W}(d, k, 1)$, that is linearizing and algebraizing. Using the preceding notations and concerning webs $\mathcal{W}(d, k, 1)$ of maximal 1-rank, the known results are essentially:

**Theorem 3.3** *Let* $\mathcal{W}(d) = \mathcal{W}(d, k, 1)$ *be a $d$-web of codimension 1 in* $(\mathbb{C}^k, 0)$ *with $k \geq 2$ whose 1-rank is maximal (i.e. $\mathrm{rk}_1 \mathcal{W}(d) = \pi(d, k, 1)$). Then P.A. Griffiths proved that $\mathcal{W}(d)$ can be algebraized if it can be linearized (cf.* [32]).

Moreover, $\mathcal{W}(d)$ can be linearized for $k \geq 2$ in cases $d \leq k$, $d = k + 1$ (*i.e.* the analogous of proposition 1 and its proof) or $d = 2k$ (cf. proposition 4). Further,

for $k = 2$, there exists a counter-example for the linearization of $\mathcal{W}(d)$ for $d = 5$, namely Bol's 5-web $\mathcal{B}$ in $(\mathbb{C}^2, 0)$ already cited;

for $k \geq 3$, there are counter-examples for the linearization of $\mathcal{W}(d)$ for $k + 1 < d < 2k$ (cf. [18]);

for $k = 3$, G. Bol proved that $\mathcal{W}(d)$ can be linearized for all $d > 5$ (cf. [10]);

for $k > 3$, S.S. Chern and P.A. Griffiths proved that $\mathcal{W}(d)$ can be linearized if $d > 2k$ with the extra assumption that $\mathcal{W}(d)$ is "normal" (cf. [18]).

Concerning the effective algebraization of linear webs $\mathcal{L}(d, k, 1)$, that is the existence and the explicit determination of the reduced non-degenerate algebraic curve $V_1 \subset \mathbb{P}^k$ of degree $d$, not necessarily irreducible and possibly singular, such that $\mathcal{L}(d, k, 1) = \mathcal{L}_{V_1}(d, k, 1)$, we refer to [Hé5].

Contrary to the first part of the preceding theorem which generalizes theorem 2 and is also a reciprocal of Abel's theorem, we establish and use in [42] a reciprocal of Reiss theorem (cf. [58,34], and also [59,3,65]).

### 3.2 Algebraization of webs $\mathcal{W}(d, 2, n)$ of maximal n-rank

This paragraph completes the preceding one and uses the same notations.

Let $\mathcal{W}(d, 2, n)$ be a $d$-web of codimension $n$ in $(\mathbb{C}^{2n}, 0)$ with maximal $n$-rank. We want to determine conditions implying that the web can be algebraized, which means that, up to a local analytical automorphism of $\mathbb{C}^{2n}$, it is algebraic (i.e. $\mathcal{W}(d, 2, n) = \mathcal{L}_{V_n}(d, 2, n)$ for a suitable reduced algebraic hypersurface $V_n$ in $\mathbb{P}^{n+1}$ of degree $d$).

It has to be noted that theorem 7 implies that any web $\mathcal{W}(d, 2, n)$ which can be algebraized necessarily has maximal $n$-rank.

Generalizing the construction of H. Poincaré presented earlier (cf. propositions 3 and 4), we will prove how to use the space $\mathcal{A}^n$ of Abelian relations of degree $n$ of the web $\mathcal{W}(d, 2, n)$ to solve the algebraization problem.

From the example $\mathcal{B} = \mathcal{B}(5, 2, 1)$ of G. Bol already mentioned, but also the three examples $\mathcal{G}_i(4, 2, 2)$ of V.V. Goldberg (cf. [24,25] and [26]) and the example $\mathcal{K}(4, 2, 2)$ of J.B. Little (cf. [51]), we conclude that assuming that the web has maximal $n$-rank is not enough to ensure that it can be algebraized.

The complete search of such examples and, more precisely, the characterization of exceptional webs $\mathcal{E}(d, 2, n)$ (i.e. of maximal $n$-rank and which cannot be algebraized) remains to be done. We will come back to it in the next paragraph.

Let $\mathcal{W}(d,2,n)$ be a $d$-web of codimension $n$ in $(\mathbb{C}^{2n},0)$ with $d \geq n+2$ with maximal $n$-rank, then

$$r_n(d) := \mathrm{rk}_n\, \mathcal{W}(d,2,n) = \binom{d-1}{n+1} \geq 1$$

so that there is at least one non trivial Abelian relation of degree $n$.

Let $\left(\gamma_i^j(F_{i_1},\ldots,F_{i_n})\right)_{1 \leq i \leq d, 1 \leq j \leq r_n(d)}$ be a basis of $\mathcal{A}^n$; since the $n$-rank of $\mathcal{W}(d,2,n)$ is maximal we have $d$ germs of maps of rank at most $n$

$$Z_i : (\mathbb{C}^{2n},0) \longrightarrow (\mathbb{P}^{r_n(d)-1}, Z_i(0))$$

defined by $Z_i(x) = [\gamma_i^1(F_{i_1},\ldots,F_{i_n}),\ldots,\gamma_i^{r_n(d)}(F_{i_1},\ldots,F_{i_n})]$ and it can be checked that the $d$ points $Z_i(x)$ depend only on $\mathcal{W}(d,2,n)$.

The *vector relation*

$$\sum_{i=1}^{d} Z_i(x).\Omega_i(x) = 0$$

necessarily holds for $x$ near $0 \in \mathbb{C}^{2n}$. Further, for $x$ near $0 \in \mathbb{C}^{2n}$ we get linear subspaces in $\mathbb{P}^{r_n(d)-1}$

$$\{Z_i(x)\} = \mathbb{P}^{d-n-2}(x)$$
$$\left\{Z_i(x), \frac{\partial Z_i}{\partial x_k}(x)\right\} = \mathbb{P}^{(n+1)(d-n-2)}(x)$$

where $\mathbb{P}^{d-n-2}(x)$ (resp. $\mathbb{P}^{(n+1)(d-n-2)}(x)$) has dimension $d-n-2$ (resp. $(n+1)(d-n-2)$).

Under the preceding hypotheses, we may even assume that

$$\{Z_i(x)\} = \{Z_{n+2}(x),\ldots,Z_d(x)\} = \mathbb{P}^{d-n-2}(x) \subset \mathbb{P}^{r_n(d)-1};$$

so that there are $(n+1)$ *independent* relations between the $Z_i(x)$ which can be presented "explicitly".

After an eventual linear change in the coordinates $(x_1,\ldots,x_{2n})$ in $\mathbb{C}^{2n}$ one can assume that

$$\delta_i(0) \neq 0 \text{ for } 1 \leq i \leq d \text{ where } \delta_i = \frac{\partial(F_{i_1},\ldots,F_{i_n})}{\partial(x_1,\ldots,x_n)}.$$

Moreover, with the help of the general position assumption, the aforementioned vectorial relation implies that, as in [19] for the case $n=2$ and in [51] for $n \geq 2$ (and through identification between Pfaff forms and vector fields) that the $d$ leaves of a web $\mathcal{W}(d, 2, n)$ of maximal $n$-rank are given for $1 \leq i \leq d$ by the *integrable system* of vector fields having the following *normal form*:

$$
(\star) \quad
\begin{cases}
X_{i_1} = \partial_{n+1} - b_i \{A_{11}\partial_1 + A_{12}\partial_2 + \cdots + A_{1n}\partial_n\} \\
X_{i_2} = \partial_{n+2} - b_i \{A_{21}\partial_1 + A_{22}\partial_2 + \cdots + A_{2n}\partial_n\} \\
\qquad \vdots \\
X_{i_n} = \partial_{2n} - b_i \{A_{n1}\partial_1 + A_{n2}\partial_2 + \cdots + A_{nn}\partial_n\}
\end{cases}
$$

where $\partial_k = \dfrac{\partial}{\partial x_k}$ for $1 \leq k \leq 2n$, and with $b_i$ and $A_{lm}$ in $\mathcal{O} = \mathbb{C}\{x_1, \ldots, x_{2n}\}$ such that the square $(n \times n)$ matrix $(A_{lm}) \in \mathcal{O}^{n^2}$ has determinant $\det(A_{lm})(0) \neq 0$. The general position assumption also *prescribes* that we have $b_i(0) \neq b_j(0)$ for $1 \leq i < j \leq d$.

Remark that the *additional* assumption concerning general position, introduced by J.B. Little in [51] (*cf.* also [19]), is satisfied here since the dimension and the codimension of the leaves of the web $\mathcal{W}(d, 2, n)$ are the same.

*Remark 3.* Using the normal form $(\star)$ and choosing suitable minors in the invertible matrix $(A_{lm})$, we can check that, for $1 \leq i \leq d$ and for $x$ near $0 \in \mathbb{C}^{2n}$, the normals $\Omega_i(x)$ of a web $\mathcal{W}(d, 2, n)$ of maximal $n$-rank with $d \geq n + 2$ belong to *a normal rational curve* $D(x)$ *in a* $n$-*plane* $\mathbb{P}^n(x) \subset \mathbb{P}^{\binom{2n}{n}-1}$ (*i.e.* $D(x)$ is, up to a linear automorphism of $\mathbb{P}^n(x)$, parameterized by $b \longmapsto [1, b, b^2, \ldots, b^n]$). This property generalizes to many "general enough" webs of codimension $n$ in $(\mathbb{C}^{kn}, 0)$ for $k \geq 2$ (*cf.* [19] for $n = 2$ and [51] for $n \geq 2$).

Moreover, the $(n + 1)$ independent relations between the $Z_i(x)$ can be written as

$$
\sum_{i=1}^{d} b_i^q(x)\delta_i(x)Z_i(x) = 0 \text{ for } 0 \leq q \leq n \qquad (R_q)
$$

for $x$ near $0 \in \mathbb{C}^{2n}$ and we have

$$
\sum_{i=1}^{d} b_i^{n+1}(0)\delta_i(0)Z_i(0) \neq 0
$$

since $b_i(0) \neq b_j(0)$ for $1 \leq i < j \leq d$.

Let $\mathcal{W}(d, 2, n)$ be a $d$-web of codimension $n$ in $(\mathbb{C}^{2n}, 0)$ *with* $d \geq n + 3$ with maximal $n$-rank, then

$$
r_n(d) = \mathrm{rk}_n \, \mathcal{W}(d, 2, n) = \binom{d - 1}{n + 1} \geq n + 2.
$$

Using the previous notations, we have a map

$$E : (\mathbb{C}^{2n}, 0) \times \mathbb{P}^1 \longrightarrow \mathbb{P}^{r_n(d)-1}$$

defined through

$$E(x, b) = \prod_{i=1}^{d} (b - b_i(x)) \cdot \sum_{i=1}^{d} \frac{\delta_i(x)}{b - b_i(x)} Z_i(x) \qquad (1)$$

$$E(x, \infty) = \sum_{i=1}^{d} b_i^{n+1}(x) \delta_i(x) Z_i(x) \qquad (2)$$

which satisfies

$$E(x, b_i(x)) = Z_i(x) \text{ for } 1 \le i \le d.$$

If we define $\Pi(x, b) = \prod_{i=1}^{d} (b - b_i(x))$, we can prove using relations $(R_0)$ and $(R_1)$ that, for $1 \le m \le n$, we have

$$\partial_{n+m}(\frac{E}{\Pi}) - b\left[\partial_1(A_{m1}\frac{E}{\Pi}) + \cdots + \partial_n(A_{mn}\frac{E}{\Pi})\right] = \sum_{i=1}^{d} \frac{X_{i_m}(b_i)\delta_i}{(b - b_i)^2} Z_i.$$

In case $n = 1$, we can always assume $X_{i_1} = \partial_2 - b_i \partial_1$ with $b_i(0) \ne b_j(0)$ for $1 \le i < j \le d$. What preceeds suggests the introduction in that case of the *unique* polynomial, up to renaming of variables and sign changes as in sect. 2.4,

$$P = \sum_{k=0}^{d} P_k b^{k-1} \in \mathcal{O}[b]$$

such that $P(x_1, x_2, b_i) = X_{i_1}(b_i)$ for $1 \le i \le d$.

Under the previous assumptions, the following results can be verified through computations (*cf.* [40] and [43]):

**Proposition 5** *For $x$ near $0 \in \mathbb{C}^{2n}$, the map $b \longmapsto E(x, b)$ gives a parameterization of a normal rational curve $E(x)$ in $\mathbb{P}^{d-n-2}(x)$ containing the $Z_i(x)$ for $1 \le i \le d$. We always have the inequality $\operatorname{rk} E \ge n+1$ and more precisely:*

*a) if $n = 1$, $\operatorname{rk} E = 2$ if and only if $\deg P \le 3$;*

*b) if $n \ge 2$, the integrability conditions of system $(\star)$ imply that $\operatorname{rk} E = n+1$.*

*Further, for $n \geq 1$ and provided that $\mathrm{rk}\, E = n+1$, then every curve $E(x)$ near $E(x_0)$ meets $\overline{E}(x_0)$ in only one point.*

With the help of this proposition, we obtain a suitable projection to a $\mathbb{P}^{n+1}$ if $\mathrm{rk}\, E = n+1$. Then, using a theorem of P.A. Griffiths already used in the proof of proposition 4, we obtain the following results (*cf.* [40] and [43] for details):

**Theorem 3.4** *Let $\mathcal{W}(d,2,n)$ be a $d$-web of codimension $n$ in $(\mathbb{C}^{2n},0)$ with maximal $n$-rank, then*

a) *if $n=1$, then $\mathcal{W}(d,2,1)$ can be algebraized if and only if $\deg P \leq 3$;*

b) *for $n \geq 2$ and $d \geq n+3$, $\mathcal{W}(d,2,n)$ can be algebraized.*

The special case $n=2$ in the theorem above has been proved by V.V. Goldberg (*cf.* [23] and [27] for details) using a completely different method based upon the important differential notion of *almost grassmannizable* web (*cf.* for instance [27]) and the following result is due to J.B. Little (*cf.* [51]):

*Every "general enough" $d$-web $\mathcal{W}(d,k,n)$ of codimension $n$ in $(\mathbb{C}^{kn},0)$ of maximal $n$-rank is almost grassmannizable as soon as $n \geq 2$ and $d > n(k-1)+2$.*

The link between the aforementioned method and the previous result of J.B. Little is the existence (*cf.* [19] and [51]) of normal forms for "general enough" webs $\mathcal{W}(d,k,n)$ of maximal $n$-rank.

## 3.3  Exceptional webs $\mathcal{E}(d,2,n)$

In this paragraph, we study particular kinds of webs $\mathcal{W}(d,2,n)$.

A $d$-web $\mathcal{E}(d,2,n)$ of codimension $n$ in $(\mathbb{C}^{2n},0)$ is called *exceptional* if it has maximal $n$-rank and it cannot be algebraized.

Theorem 9 implies that such $\mathcal{E}(d,2,n)$, if they exist, necessarily correspond to $n=1$ and $d \geq 5$ (since otherwise $\deg P \leq 3$) or $n \geq 2$ and $3 \leq d \leq n+2$ and in that case we have $\mathrm{rk}_n\, \mathcal{E}(n+2,2,n) = 1$.

The only known examples of exceptional webs have already been cited, and will be recalled again.

One of these suggests a brief presentation of observations concerning *polylogarithmic* webs in $\mathbb{C}^2$. Indeed, the study of such webs deserves to be carried on.

Using the power series $\sum_{n \geq 1} \dfrac{z^n}{n^p}$, we can construct for all $p \geq 1$ the *polylog-arithms* $\phi_p$ *of order* $p$. These satisfy particular functional relations which can be viewed as Abelian relations for suitable webs in $\mathbb{C}^2$.

More generally, the theory of polylogarithms is related to numerous domains (homology of the linear group, algebraic K-theory, hyperbolic geometry, projective configurations, *etc.*) for which we refer, as for precise definitions of $\phi_p$, to the classic book [46] of L. Lewin and more or less recent articles and reports (*cf.* notably [36,47,56,14]).

The logarithm $\phi_1(z) = \log z$ satisfies the functional relation with 3 terms

$$\phi_1(x) - \phi_1(y) + \phi_1(\frac{y}{x}) = 0$$

using the appropriate determinations; it gives rise to the 3-web $\mathcal{P}log(3)$ in $\mathbb{C}^2$ defined by the functions $x$, $y$ and $y/x$. This web has already been presented (*cf.* sect. 2.3)) and we saw that $\mathcal{P}log(3) = \mathcal{P}(3)$ is the 3-web in $\check{\mathbb{P}}^2$ formed by the 3 pencils of lines in $\check{\mathbb{P}}^2$ with vertices $[0,0,1]$, $[0,1,0]$ and $[1,0,0]$.

The 3-web $\mathcal{P}log(3)$ in $\mathbb{C}^2$ admits $S = \{xy = 0\}$ as singular locus, where the singular locus is defined as the set of points in $\mathbb{C}^2$ at which the 3 foliations defining $\mathcal{P}log(3)$ are either singular or do not intersect transversally.

In fact, the main object associated with the 3-web $\mathcal{P}log(3)$ in $\mathbb{C}^2$ is its differential linear system:

$$\mathcal{R}(3) \quad \begin{cases} \partial_x(f_1) = 0 \\ \partial_y(f_2) = 0 \\ (x\partial_x + y\partial_y)(f_3) = 0 \\ \partial_x(f_1 + f_2 + f_3) = 0 \\ \partial_y(f_1 + f_2 + f_3) = 0 \end{cases}$$

and the associated left $\mathcal{D}$-module will also be called $\mathcal{R}(3)$ (*cf.* sect. 2.1) and for instance [31,45] concerning the study of this object and the following ones).

It can be proven that $\mathcal{R}(3)$ is holonomic and that its characteristic cycle is

$$[\mathrm{Car}\,\mathcal{R}(3)] = 4.T^*_{\mathbb{C}^2}\mathbb{C}^2 + 1.T^*_{\{x=0\}}\mathbb{C}^2 + 1.T^*_{\{y=0\}}\mathbb{C}^2.$$

The *index* of the left $\mathcal{D}$-module $\mathcal{R}(3)$ defined, for $z = (x,y) \in \mathbb{C}^2$ through

$$\chi_z(\mathcal{R}(3)) := \sum_{i \geq 0}(-1)^i \dim_{\mathbb{C}} \mathcal{E}xt^i_{\mathcal{D}}(\mathcal{R}(3), \mathcal{O})_z$$

is a good substitute for the rank in the study of singularities of the corresponding web. Indeed, one can check through Kashiwara's index formula that

$$\chi_z(\mathcal{R}(3)) = \begin{cases} 4 & \text{if } z \notin S \\ 3 & \text{if } z \in S - \{(0,0)\} \\ 2 & \text{if } z = (0,0) \end{cases}$$

which *differenciates* the nature of singular points of $\mathcal{P}log(3)$; moreover, the $\mathbb{C}$-vector spaces $\mathcal{H}om_D(\mathcal{R}(3), \mathcal{O})$ and $\mathrm{Sol}\,\mathcal{R}(3)$ can be identified; this proves that the web $\mathcal{P}log(3)$ is hexagonal for $z \notin S$ since in this case the "upper" solutions do not occur and then

$$\chi_z(\mathcal{R}(3)) = \dim_{\mathbb{C}} \mathcal{E}xt_D^0(\mathcal{R}(3), \mathcal{O})_z = \mathrm{rk}_z\,\mathcal{P}log(3) + 3 = 4.$$

The dilogarithm $\phi_2$ satisfies the functional relation with 5 terms

$$\phi_2(x) - \phi_2(y) + \phi_2(\frac{y}{x}) - \phi_2(\frac{1-y}{1-x}) + \phi_2(\frac{x}{y} \cdot \frac{1-y}{1-x}) = 0$$

and gives rise to the 5-web $\mathcal{P}log(5)$ in $\mathbb{C}^2$. Up to a projective transformation, it can be checked that

$$\mathcal{P}log(5) = \mathcal{B} = \mathcal{B}(5,2,1)$$

where $\mathcal{B}$ stands for Bol's 5-web in $\mathbb{C}^2$ (*cf.* sect. 2.2)); the vertices of the four pencils of lines are in this case $[0,0,1]$, $[0,1,0]$, $[1,0,0]$ and $[1,1,1]$.

We have $\mathrm{rk}\,\mathcal{P}log(5) = 6$ (*cf.* [11]); a *basis* of Abelian relations of $\mathcal{P}log(5) = \mathcal{B}$ is given by 5 relations on 3 terms induced by $\phi_1$ (remark that $\mathcal{B}$ is hexagonal (*cf.* theorem 4)) completed with the relation on 5 terms *precisely* satisfied by $\phi_2$.

Since $\mathcal{P}log(5)$ is of rank 6 and cannot be linearized, it is an exceptional web $\mathcal{E}(5,2,1)$. A thorough study of the nature of its singularities, as has been done for $\mathcal{P}log(3)$ would be very interesting.

The trilogarithm $\phi_3$ satisfies the functional relation on 9 terms

$$2\phi_3(x) + 2\phi_3(y) - \phi_3(\frac{x}{y}) + 2\phi_3(\frac{1-x}{1-y}) + 2\phi_3(\frac{x}{y} \cdot \frac{1-y}{1-x}) \tag{3}$$

$$+2\phi_3(x \cdot \frac{1-y}{x-1}) + 2\phi_3(y \cdot \frac{1-x}{y-1}) - \phi_3(xy) - \phi_3(\frac{x}{y} \cdot \frac{(1-y)^2}{(1-x)^2}) = 2\zeta(3) \tag{4}$$

and gives rise to the 9-web $\mathcal{P}log(9)$ in $\mathbb{C}^2$. This web *cannot be linearized* since one of its extracted webs is $\mathcal{P}log(5) = \mathcal{B}$.

The study of the singularities of $\mathcal{P}log(9)$ in the preceding sense is interesting, but above all this web seems to be a good candidate as an exceptional web $\mathcal{E}(9,2,1)$.

In other respects, most of the detailed study of the geometrical hierarchy of algebraic curves constituting leaves of the polylogarithmic webs $\mathcal{P}log(d)$ in $\mathbb{C}^2$ already envisioned remains to be done. Indeed, this could be useful to exhibit functional relations for $\phi_p$ for $p \geq 3$.

For $n \geq 2$, the only known examples of exceptional webs are of the form $\mathcal{E}(4,2,2)$ so that their 2-rank is 1. These are on one hand the three $\mathcal{G}_i(4,2,2)$ of V.V. Goldberg (cf. [27] for details) which are explicit and belong to two different classes, and on the other hand, the web $\mathcal{K}(4,2,2)$ of J.B. Little (cf. [51]) which is not explicit, constructed from a smooth quartic in $\mathbb{P}^3$ using general properties of the 0-cycles in a $K3$ surface proved by D. Mumford and A.A. Roitman.

Other invariants of webs $\mathcal{W}(d,k,n)$ will be presented for $n \geq 2$ in the next paragraph; these should give a better grasp on $\mathcal{E}(d,2,n)$, notably concerning the examples of V.V. Goldberg and J.B. Little.

For $n \geq 1$ and in view of the known examples of exceptional webs, it should be worth investigating webs $\mathcal{W}(d,k,n)$ through their Abelian relations of degree $n$ of the particular form

$$\sum_{i=1}^{d} m_i \mathbf{g}(F_{i_1}, \ldots, F_{i_n}) dF_{i_1} \wedge \cdots \wedge dF_{i_n} = 0$$

with one $\mathbf{g} \in \mathbb{C}\{z_1, \ldots, z_n\}$ and integers $m_i \in \mathbb{Z}$.

### 3.4   The complex of Abelian relations of a web $\mathcal{W}(d,2,n)$

We continue with the notations of the previous paragraphs.

Let $\mathcal{W}(d,2,n)$ be a $d$-web of codimension $n$ in $(\mathbb{C}^{2n}, 0)$ defined by elements $F_{i_m} \in \mathcal{O} = \mathbb{C}\{x_1, \ldots, x_{2n}\}$ with $F_{i_m}(0) = 0$ for $1 \leq i \leq d$ and $1 \leq m \leq n$.

For $1 \leq p \leq n$, $1 \leq i \leq d$ and for any multi-index $I_i = \{i_{m_1}, \ldots, i_{m_p}\}$ of length $|I_i| = p$, of elements of $\{i_1, \ldots, i_n\}$ in increasing order we let $dF_{I_i} = \bigwedge_{j=1}^{p} dF_{i_{m_j}}$.

Related to an idea of P.A. Griffiths (*cf.* [33]), we set

$$A^p = \left\{ \left( \alpha_{I_i}(F_{i_1}, \dots, F_{i_n}) \right)_{1 \le i \le d, |I_i| = p} \in \mathcal{O}^{\binom{n}{p} \cdot d} \text{ such that} \right.$$

$$\alpha_{I_i} \in \mathbb{C}\{z\} = \mathbb{C}\{z_1, \dots, z_n\} \text{ and } \sum_{1 \le i \le d, |I_i| = p} \alpha_{I_i}(F_{i_1}, \dots, F_{i_n}) \, dF_{I_i} = 0 \left. \right\}$$

the $\mathbb{C}$-vector space of *Abelian relations of degree $p$* of the web $\mathcal{W}(d, 2, n)$; for $p = n$, these are the Abelian relations of maximal degree $n$ of the web $\mathcal{W}(d, 2, n)$.

The exterior derivative induces a *complex* $(A^\bullet, \delta)$ of $\mathbb{C}$-vector spaces (*i.e.* $\delta \circ \delta = 0$).

For $1 \le i \le d$, let $(X_{i_m})$ be a family of $n$ vector fields in $(\mathbb{C}^{2n}, 0)$ associated with the leaves of $\mathcal{W}(d, 2, n)$. We consider the following linear differential system:

$$\mathcal{R} \quad \begin{cases} X_{i_m}(f_i) = 0 \text{ for } 1 \le i \le d \text{ and } 1 \le m \le n \\ \partial_k(f_1 + \cdots + f_d) = 0 \text{ for } 1 \le k \le 2n \end{cases}$$

The $\mathbb{C}$-vector space of solutions $(f_1, \dots, f_d) \in \mathcal{O}^d$ of the linear differential system $\mathcal{R}$ will be denoted by $A^0 := \mathrm{Sol}\,\mathcal{R}$.

The usual differential induces a $\mathbb{C}$-linear map

$$\delta : A^0 \longrightarrow A^1.$$

Indeed, any element $f_i \in \mathcal{O}$ satisfying $X_{i_m}(f_i) = 0$ for $1 \le m \le n$ can be written as $f_i = \alpha_i(F_{i_1}, \dots, F_{i_n})$ for a suitable $\alpha_i \in \mathbb{C}\{z\}$ and through the definition of $A^1$ we get

$$\delta(f_1, \dots, f_d) := \left( \frac{\partial \alpha_i}{\partial z_1}(F_{i_1}, \dots, F_{i_n}), \dots, \frac{\partial \alpha_i}{\partial z_n}(F_{i_1}, \dots, F_{i_n}) \right) \in A^1$$

for $(f_1, \dots, f_d) \in A^0$.

The complex $(A^\bullet, \delta)$ extended to $p = 0$ is called the *complex of Abelian relations* of the web $\mathcal{W}(d, 2, n)$.

The preceding definitions yield that

$$A^0 = \left\{ (\alpha_i(F_{i_1}, \dots, F_{i_n})) \in \mathcal{O}^d \mid \alpha_i \in \mathbb{C}\{z\} \text{ and } \sum_{i=1}^d \alpha_i(F_{i_1}, \dots, F_{i_n}) = cst \right\}$$

where, as in the case of Abelian relations of degree $p$ of the web $\mathcal{W}(d, 2, n)$, the function $\alpha_i(F_{i_1}, \dots, F_{i_n})$ is constant along the leaf indexed by $i$.

The elements of the $\mathbb{C}$-vector space $\mathcal{A}^0$ can be considered as *"functions"* over the $d$-web $\mathcal{W}(d, 2, n)$, and, for $1 \leq p \leq n$ the elements of $\mathcal{A}^p$ play the role of *"p-forms"*.

Further, it can be checked through the usual Poincaré lemma about differential forms, built upon the $dz_i$ in $\mathbb{C}\{z\}$, that we have an exact sequence of $\mathbb{C}$-vector spaces

$$0 \longrightarrow \mathbb{C}^d \longrightarrow \mathcal{A}^0 \xrightarrow{\delta} \mathcal{A}^1 \xrightarrow{\delta} \mathcal{A}^2.$$

Adapting the Poincaré-Blaschke method (*cf.* sect. 2.1)), we can bound the dimension of $\mathcal{A}^p$ for $1 \leq p \leq n$.

Let $1 \leq p \leq n$ and let $\left(\alpha_{I_i}^j(F_{i_1}, \ldots, F_{i_n})\right)_{1 \leq i \leq d, |I_i|=p, 1 \leq j \leq r_p}$ be an independent family of elements of $\mathcal{A}^p$. We have $\binom{n}{p}.d$ germs of maps of rank at most $n$

$$Z_{I_i} : (\mathbb{C}^{2n}, 0) \longrightarrow (\mathbb{C}^{r_p}, Z_{I_i}(0))$$

defined, if we omit the dependency on the variables $x = (x_1, \ldots, x_{2n})$ by $Z_{I_i}(x) = \left(\alpha_{I_i}^1(F_{i_1}, \ldots, F_{i_n}), \ldots, \alpha_{I_i}^{r_p}(F_{i_1}, \ldots, F_{i_n})\right)$. These germs satisfy, for $x$ near $0 \in \mathbb{C}^{2n}$, the dependence relation:

$$\sum_{1 \leq i \leq d, |I_i|=p} Z_{I_i}(x) dF_{I_i}(x) = 0.$$

For instance, if $p = 1$ and omitting the dependency on $x$ (*cf.* notations of 2.1), we get

$$\{Z_{I_i}(x)\} = \{Z_{I_1}; \ldots; Z_{I_d}\} = \{Z_{1_1}, \ldots, Z_{1_n}; \ldots; Z_{d_1}, \ldots, Z_{d_n}\}$$

and $\sum_{i=1}^{d} Z_{i_1} dF_{i_1} + \cdots + Z_{i_n} dF_{i_n} = 0$, which gives $2n$ relations between the $Z_{i_m}$.

Using the general position assumption, one may assume that

$$\{Z_{I_i}(x)\} = \{Z_{3_1}, \ldots, Z_{3_n}; \ldots; Z_{d_1}, \ldots, Z_{d_n}\}$$

if we choose $F_{1_1} = x_1, \ldots, F_{1_n} = x_n$ and $F_{2_1} = x_{n+1}, \ldots, F_{2_n} = x_{2n}$. Then, derivation modulo $\{Z_{I_i}\}$ of the preceding dependence relation and another use of the general position assumption yields that

$$\{Z_{I_i}(x), \partial_k(Z_{I_i})(x)\} = \{Z_{3_1}, \ldots, Z_{3_n}; \ldots; Z_{d_1}, \ldots, Z_{d_n}; \partial_1(Z_{4_1}), \ldots, \partial_1(Z_{4_n}); \ldots;$$
$$\partial_1(Z_{d_1}), \ldots, \partial_1(Z_{d_n}); \ldots; \partial_n(Z_{4_1}), \ldots, \partial_n(Z_{4_n}); \ldots; \partial_n(Z_{d_1}), \ldots, \partial_n(Z_{d_n})\}.$$

After enough derivations, this procedure comes to an end and since we started with an independent family of $\mathcal{A}^1$, we finally obtain the bound

$$r_1 \leq n \left[ \{d-2\} + \binom{n}{1}\{d-3\} + \binom{n+1}{2}\{d-4\} + \cdots \right]$$

where the $\binom{n+q-1}{q}$ correspond to the $q$-th order of derivation. This can also be rewritten as

$$r_1 \leq \binom{n}{1} \cdot \binom{d+n-2}{n+1}.$$

More generally, for $1 \leq p \leq n$ and using the same method, we come to the bounds

$$r_p \leq \binom{n}{p} \cdot \binom{d+n-p-1}{n+1}.$$

In the $C^\infty$ setting, we can also define a complex $(\mathcal{A}^\bullet, \delta)$ with the same properties and the above method, suitably adapted, should give the same bounds.

Incidentally, if we generalize the method presented at the end of sect. 2.1 one can, using recurrence on $d$, prove that the left module $\mathcal{R}$ associated to the linear differential system with the same name is holonomic (in fact it is a connection) and that its multiplicity $\operatorname{mult}\mathcal{R}$, that is the dimension of the space of its solutions is bounded by

$$\operatorname{mult}\mathcal{R} = \dim_{\mathbb{C}} \mathcal{A}^0 \leq \binom{d+n-1}{n+1} + 1.$$

The preceding results can be gathered as the following theorem:

**Theorem 3.5** *Let $\mathcal{W}(d,2,n)$ be a $d$-web of codimension $n$ in $(\mathbb{C}^{2n}, 0)$, we have a complex $(\mathcal{A}^\bullet, \delta)$ of finite dimensional $\mathbb{C}$-vector spaces with the following bounds:*

$$r_0 := \dim_{\mathbb{C}} \mathcal{A}^0 \leq \binom{d+n-1}{n+1} + 1$$

*and, for $1 \leq p \leq n$,*

$$r_p := \dim_{\mathbb{C}} \mathcal{A}^p \leq \binom{n}{p} \cdot \binom{d+n-p-1}{n+1}.$$

The integers $r_p$ defined above for $0 \leq p \leq n$ are analytical invariants of the web $\mathcal{W}(d, 2, n)$ which do not depend on the choice of the $F_{i_m}$; for $0 \leq p \leq n$, $r_p = \mathrm{rk}_p \mathcal{W}(d, 2, n)$ is called the *p-rank* of the web.

The bound for $r_0$ obtained above appears in [27] for different webs $\mathcal{W}(d, 2, n)$ sharing special properties, namely almost grassmannizable ones.

The 1-rank can be used, for instance, in order to characterize 3-webs of codimension $n$ in $(\mathbb{C}^{2n}, 0)$ which are *parallelizable*. Indeed, with the preceding notations, if $\mathcal{W}(3, 2, n)$ has *maximal* 1-rank (*i.e.* $r_1 = n$) we can use a basis

$$\left(\alpha_{i_m}^j (F_{i_1}, \ldots, F_{i_n})\right)_{1 \leq i \leq 3, 1 \leq m \leq n, 1 \leq j \leq r_1}$$

of $\mathcal{A}^1 = \mathcal{A}^1\left(\mathcal{W}(3, 2, n)\right)$ to build an isomorphism $\phi : (\mathbb{C}^{2n}, 0) \longrightarrow (\mathbb{C}^{2n}, 0)$ which *parallelizes* $\mathcal{W}(3, 2, n)$. It is enough to set

$$\phi(x) = \left(\widetilde{\alpha}_i^j (F_{i_1}(x), \ldots, F_{i_n}(x))\right)_{1 \leq i \leq 2, 1 \leq j \leq r_1}$$

where $\sum_{i=1}^{3} \widetilde{\alpha}_i^j (F_{i_1}, \ldots, F_{i_n}) = 0$ with $\delta\left(\widetilde{\alpha}_i^j (F_{i_1}, \ldots, F_{i_n})\right) = \left(\alpha_{i_m}^j (F_{i_1}, \ldots, F_{i_n})\right)$ since $\mathcal{A}^2\left(\mathcal{W}(3, 2, n)\right) = 0$ and using the following identities:

$$\{Z_{i_1}, \ldots, Z_{i_n}\} = \{Z_{3_1}, \ldots, Z_{3_n}\}$$
$$= \{Z_{i_1}, \ldots, Z_{i_n}, \partial_k(Z_{i_1}), \ldots, \partial_k(Z_{i_n}), \ldots\} = \mathbb{C}^n.$$

As a consequence, we obtain the following generalization of proposition 1:

**Proposition 6** *Let $\mathcal{W}(3, 2, n)$ be a 3-web of codimension $n$ in $(\mathbb{C}^{2n}, 0)$, the following conditions are equivalent:*

(i) $\mathrm{rk}_1 \mathcal{W}(3, 2, n) = n$ *(i.e. $\mathcal{W}(3, 2, n)$ has maximal 1-rank);*

(ii) *Up to a local analytical isomorphism of $(\mathbb{C}^{2n}, 0)$, the 3-web $\mathcal{W}(3, 2, n)$ is defined by the following families of n-planes in $\mathbb{C}^{2n}$:*

$$\{x_1 = cst, \ldots, x_n = cst\}, \{x_{n+1} = cst, \ldots, x_{2n} = cst\}$$
$$\text{and } \{x_1 + x_{n+1} = cst, \ldots, x_n + x_{2n} = cst\}.$$

This proposition can be compared with the following corollary of proposition 4 in case $k = 2$:

*A web $\mathcal{W}(n+3, 2, n)$ can be algebraized if and only if its n-rank is maximal (i.e. $\mathrm{rk}_n \mathcal{W}(n + 3, 2, n) = n + 2$).*

The formula

$$d - 1 = \sum_{p=0}^{n} (-1)^p \binom{n}{p} \cdot \binom{d+n-p-1}{n+1}$$

shows that the bounds on $r_p = \mathrm{rk}_p \, \mathcal{W}(d,2,n)$ presented above are *linked*; this is not a surprise in view of the aforementioned and the existence of the complex $(\mathcal{A}^\bullet, \delta)$. Moreover, the formula presented above has some incidence on the cohomology of the complex $(\mathcal{A}^\bullet, \delta)$, notably concerning its Euler characteristic.

The preceding results generalize to webs $\mathcal{W}(d,k,n)$ and make remark 2 (or at least try to) more convincing; particularly, the bounds given above for the $r_p$ for $1 \le p \le n$ are particular instances of generalized Castelnuovo numbers $\pi_p(d,k,n)$, namely $\pi_p(d,2,n) := \binom{n}{p} \cdot \binom{d+n-p-1}{n+1}$.

*Remark 4.* Let $V_n \subset \mathbb{P}^{n+1}$ be a reduced algebraic hypersurface of degree $d$. Let $(\mathcal{B}^\bullet, d) := (H^0(V_n, \omega_{V_n}^\bullet), d)$ be the complex of Abelian differential forms on $V_n$, that is the complex induced by differentials on the global sections of Barlet $p$-sheaves over $V_n$ for $0 \le p \le n$ (*cf.* [6]).

It can be proven, generalizing results of sect. 3.1 that there exists an *injective morphism* between complexes

$$A^\bullet : (\mathcal{B}^\bullet, d) \longrightarrow (\mathcal{A}^\bullet, \delta)$$

which is one-to-one at least in maximal degree $n$. Moreover, for $0 \le p \le n$, one can construct algebraic hypersurfaces $V_n \subset \mathbb{P}^{n+1}$ of degree $d$, singular if $p \ne n$, such that the dimension of $\mathcal{B}^p$ coincide with the bounds for $r_p$; this proves the optimality of the bounds given in theorem 10.

This way, we obtain, through theorem 10 and the morphism $A^\bullet$ between complexes, *optimal bounds for the dimension of $\mathcal{B}^p = H^0(V_n, \omega_{V_n}^p)$ for $0 \le p \le n$.*

These results generalize to algebraic varieties $V_n$ in $\mathbb{P}^{n+k-1}$ of dimension $n$ and degree $d$ which are general enough as defined in sect 3.1.

## Acknowledgments

The present work was initially supposed to be published in French. The author thanks his colleague, Gilles Robert, for the translation.

# References

1. N.H. ABEL, *Mémoire sur une propriété générale d'une classe très étendue de fonctions transcendantes*, Œuvres complètes, Tome premier, Grøndahl & Søn, Chritiania, (1881), 145-211.

2. M.A. AKIVIS, *Differential geometry of webs*, J. Soviet Math. **29** (1985), 1631-1647.

3. M.A. AKIVIS & V.V. GOLDBERG, *Projective Differential Geometry of Submani-folds*, North-Holland, Amsterdam, (1993).

4. M.A. AKIVIS & A.M. SHELEKOV, *Geometry and Algebra of Multidimensional Three-Webs*, Kluwer, Dordrecht, 1992.

5. V. ARNOLD, *Chapitres supplémentaires de la théorie des équations différentielles ordinaires*, MIR, Moscou, (1980).

6. D. BARLET, *Le faisceau $\omega_X^{\bullet}$ sur un espace analytique $X$ de dimension pure*, in Fonctions de Plusieurs Variables Complexes III, Séminaire F. Norguet, Lect. Notes Math. **670**, Springer, Berlin, (1978), 187-204.

7. A. BEAUVILLE, *Géométrie des tissus (d'après S.S. Chern et P.A. Griffiths)*, Séminaire Bourbaki, exposé 531 (février 1979), Lect. Notes Math. **770**, Springer, Berlin, 1980, 103–119.

8. W. BLASCHKE, *Einführung in die Geometrie der Waben*, Birkhäuser, Basel, 1955.

9. W. BLASCHKE, *Geometría de los tejidos*, Seminario Matemático de Barcelona, 1954.

10. W. BLASCHKE & G. BOL, *Geometrie der Gewebe*, Springer, Berlin, 1938.

11. G. BOL, *Über ein bemerkenswertes Fünfgewebe in der Ebene*, Abh. Hamburg **11** (1936), 387-393.

12. E. CARTAN, *Sur les variétés à connexion projective*, Bull. Soc. Math. France **52** (1924), 205-241.

13. G. CASTELNUOVO, *Ricerche di Geometria sulle curve algebriche*, Atti R. Accad. Sci. Torino **24** (1889), 346-373.

14. J.-L. CATHELINEAU, *Homologie du groupe linéaire et polylogarithmes (d'après A.B. Goncharov et d'autres)*, Séminaire Bourbaki, exposé 772 (juin 1993), Astérisque **216** (1993), 311-341.

15. D. CERVEAU, *Théorèmes de type Fuchs pour les tissus feuilletés*, in Complex analytic methods in dynamical systems (Rio de Janeiro, 1992), Astérisque **222** (1994), 49-92.

16. S.S. CHERN, *Abzählungen für Gewebe*, Abh. Hamburg **11** (1936), 163-170.

17. S.S. CHERN, *Web Geometry*, Bull. Amer. Math. Soc. **6** (1982), 1-8.

18. S.S. CHERN & P.A. GRIFFITHS, *Abel's Theorem and Webs*, Jahresber. Deutsch. Math.-Verein. **80** (1978), 13-110 and *Corrections and Addenda to Our Paper : Abel's Theorem and Webs*, Jahresber. Deutsch. Math.-Verein. **83** (1981), 78-83.

19. S.S. CHERN & P.A. GRIFFITHS, *An Inequality for the Rank of a Web and Webs of Maximum Rank*, Ann. Scuola Norm. Sup. Pisa **5** (1978), 539-557.

20. D.B. DAMIANO, *Webs and Characteristic Forms on Grassmann Manifolds*, Amer. J. Math. **105** (1983), 1325-1345.

21. G. DARBOUX, *Leçons sur la théorie générale des surfaces*, Livre I, 2-ième édition, Gauthier-Villars, Paris, (1914).

22. E. GHYS, *Flots transversalement affines et tissus feuilletés*, Supplément Bull. Soc. Math. France, Mémoire 46, **119** (1991), 123-150.

23. V.V. GOLDBERG, *Tissus de codimension r et de r-rang maximum*, C. R. Acad. Sc. Paris **297** (1983), 339-342.

24. V.V. GOLDBERG, *4-tissus isoclines exceptionnels de codimension deux et de 2-rang maximal*, C. R. Acad. Sc. Paris **301** (1985), 593-596.

25. V.V. GOLDBERG, *Isoclinic webs $W(4,2,2)$ of maximum 2-rank*, in Differential Geometry - Peñiscola 1985, Lect. Notes Math. **1209**, Springer, Berlin, (1986), 168-183.

26. V.V. GOLDBERG, *Nonisoclonic 2-codimensional 4-webs of maximum 2-rank*, Proc. Amer. Math. Soc. **100** (1987), 701-708.

27. V.V. GOLDBERG, *Theory of Multicodimensional $(n + 1)$-Webs*, Kluwer, Dordrecht, (1988).

28. V.V. GOLDBERG, *On the linearizability condition for a three-web on a two-dimensional manifold*, in Differential Geometry, F.J. Carreras, O. Gil-Medrano and A.M. Naveira (Eds), Lect. Notes Math. **1410**, Springer, Berlin, (1989), 223-239.

29. V.V. GOLDBERG, *Curvilinear 4-webs with equal curvature forms of its 3-subwebs*, in Webs and Quasigroups, Tver State Univ., Russia, (1993), 9-19.

30. V.V. GOLDBERG, *Special classes of curvilinear 4-webs with equal curvature forms of their 3-subwebs*, in Webs and Quasigroups, Tver State Univ., Russia, (1996), 24-39.

31. M. GRANGER & P. MAISONOBE, *A basic course on differential modules*, in $\mathcal{D}$-modules cohérents et holonomes, Travaux en cours, 45, Hermann, Paris, (1993), 103-168.

32. P.A. GRIFFITHS, *Variations on a Theorem of Abel*, Invent. Math. **35** (1976), 321-390.

33. P.A. GRIFFITHS, *On Abel's Differential Equations*, Algebraic Geometry, The Johns Hopkins Centennial Lectures, Ed. J.-I. Igusa (1977), 26-51.

34. P.A. GRIFFITHS & J. HARRIS, *Principles of Algebraic Geometry*, John Wiley & Sons, New York, (1978).

35. T.-H. GRONWALL, *Sur les équations entre trois variables représentables par des nomogrammes à points alignés*, J. de Liouville **8** (1912), 59-102.

36. R. HAIN & R. MACPHERSON, *Higher Logarithms*, Ill. J. Math. **34** (1990), 392-475.

37. J. HARRIS, *A Bound on the Geometric Genus of Projective Varieties*, Ann. Scuola Norm. Sup. Pisa **8** (1981), 35-68.

38. A. HÉNAUT, *$\mathcal{D}$-modules et géometrie des tissus de $\mathbb{C}^2$*, Math. Scand. **66** (1990), 161-172.

39. A. HÉNAUT, *Sur la linéarisation des tissus de $\mathbb{C}^2$*, Topology **32** (1993), 531-542.

40. A. HÉNAUT, *Caractérisation des tissus de $\mathbb{C}^2$ dont le rang est maximal et qui sont linéarisables*, Compositio Math. **94** (1994), 247-268.

41. A. HÉNAUT, *Systèmes différentiels, nombre de Castelnuovo et rang des tissus de $\mathbb{C}^n$*, Publ. R.I.M.S., Kyoto Univ. **31** (1995), 703-720.

42. A. HÉNAUT, *Tissus linéaires et théorèmes d'algébrisation de type Abel-inverse et Reiss-inverse*, Geom. Dedicata **65** (1997), 89-101.

43. A. HÉNAUT, *Sur l'algébrisation des tissus de codimension n de $\mathbb{C}^{2n}$*, Ann. scient. Éc. Norm. Sup. **31** (1998), 131-143.

44. J.-L. JOLY, G. MÉTIVIER & J. RAUCH, *Resonant one dimensional nonlinear geometric optics*, Journal Funct. Analysis **114** (1993), 106-231.

45. M. KASHIWARA, *Index Theorem for a Maximally Overdetermined System of Linear Differential Equations*, Proc. Japan Acad. **49** (1973), 803-804.

46. L. LEWIN, *Polylogarithms and Associated Functions*, North-Holland, New York, 1981.

47. L. LEWIN, *Structural Properties of Polylogarithms*, Mathematical Surveys and Monographs, vol.37, Amer. Math. Soc., 1991.

48. S. LIE, *Bestimmung aller Flächen, die in mehrfacher Weise durch Translationsbewegung einer Kurve erzeugt werden*, Arch. for Math. Nat. **7** (1882), 155-176.

49. R. LIOUVILLE, *Sur les invariants de certaines équations différentielles et sur leurs applications*, J. Ec. Polytechnique **59** (1889), 7-76.

50. J.B. LITTLE, *Translation manifolds and the converse of Abel's theorem*, Compositio Math. **49** (1983), 147-171.

51. J.B. LITTLE, *On Webs of Maximum Rank*, Geom. Dedicata **31** (1989), 19-35.

52. G. MIGNARD, *Rang et courbure des 3-tissus du plan et applications aux équations différentielles*, Thèse de Doctorat, Université Bordeaux I, janvier 1999.

53. G. MIGNARD, *Rang et courbure des 3-tissus de* $\mathbb{C}^2$, C. R. Acad. Sc . Paris **329** (1999), 629-632.

54. I. NAKAI, *Curvature of curvilinear 4-webs and pencils of one forms : Variation on a theorem of Poincaré, Mayrhofer and Reidemeister*, Comment. Math. Helv. **73** (1998), 177-205.

55. F. NORGUET, *Sur l'espace des cycles analytiques compacts d'un espace analytique complexe réduit*, in Fonctions analytiques de plusieurs variables et analyse complexe, Colloque international du C.N.R.S.- Paris 1972, Gauthier-Villars, Paris, (1974), 159-171.

56. J. OESTERLÉ, *Polylogarithmes*, Séminaire Bourbaki, exposé 762 (novembre 1992), Astérisque **216** (1993), 49-67.

57. H. POINCARÉ, *Sur les surfaces de translation et les fonctions abéliennes*, Bull. Soc. Math. France **29** (1901), 61-86.

58. M. REISS, *Mémoire sur les propriétés générales des courbes algébriques etc.*, Corresp. Math. et Phys. de Quetelet **9** (1837), 249-308.

59. B. SEGRE, *Sui teoremi di Bézout, Jacobi e Reiss*, Ann. Mat. Pura Appl. **26** (1947), 1-26.

60. R. THOM , *Philosophie de la singularité*, in Apologie du Logos, Hachette, Paris, 1990.

61. G. THOMSEN, *Un teorema topologico sulle schiere di curve e una caratterizzazione geometrica delle superficie isotermo-asintotiche*, Boll. Un. Mat. Ital. Bologna. **6** (1927), 80-85.

62. A. TRESSE, *Sur les invariants différentiels des groupes continus de transformation*, Acta Math. **18** (1894), 1-88.

63. A. TRESSE, *Détermination des Invariants ponctuels de l'Équation différentielle ordinaire du second ordre :* $y'' = \omega(x, y, y')$, Preisschr. der Fürst. Jablon. Ges., Breitkopf & Härtel, Leipzig, (1896).

64. W. WIRTINGER, *Lie's Translationsmannigfaltigkeiten und Abel'sche Integrale*, Monasth. Math. Phys. **46** (1938), 384-431.

65. J.A. WOOD, *A simple criterion for local hypersurfaces to be algebraic*, Duke Math. J. **51** (1984), 235-237.

# WEBS AND CURVATURE

PÉTER T. NAGY

*Department of Geometry, Lajos Kossuth University, H-4010 DEBRECEN, P.O.B. 12., Hungary, nagypeti@math.klte.hu*

This paper is devoted to the differential geometric study of 3-webs. After introducing the canonical Chern connection of a 3-web we give a full global description of left invariant webs on Lie groups. There is given a proof of Blaschke's theorem about the characterization of the hexagonal local closure condition for 3-webs on 2-dimensional manifolds by the vanishing of their curvature. We treat the theory of M. A. Akivis on 3-webs on higher dimensional manifolds: the curvature identities which are equivalent to the classical local closure conditions of Reidemeister, Bol and Moufang for 3-webs are investigated and there is given a study of transversally geodesic and torsionless 3-webs. The relations between local closure conditions and curvature identities are interpreted as weak-associative identities of binary multiplications and the corresponding infinitesimal identities of the commutator and associator operations of their tangent algebras

## 1   3-webs and $(h, j)$-structures

A 3-web is a triple of half codimensional foliations in general position. The tangent distributions of these foliations determine a pair of complementary subbundles together with the corresponding horizontal and vertical projections and an almost product structure interchanging these two distributions. We call the differential geometric structure defined by these tensorfields a $(h, j)$-structure. The tangent distributions of the foliations of the 3-web can be expressed with the help of the tensorfields of the corresponding $(h, j)$-structure. One can associate with a $(h, j)$-structure a canonical linear connection, the so called Chern connection, which satisfies a natural assumption for the torsion tensorfield and has the following property: the parallel translation with respect to this connection preserves the tensorfields of the $(h, j)$-structure. The basic tool of the study of $(h, j)$-structures and 3-webs are the Chern connection and its differential geometric invariants: torsion and curvature tensorfields and their covariant derivatives. The integrability of the distributions of the $(h, j)$-structure can be expressed by the vanishing of some components of the torsion tensorfield. The leaves of the foliations of the corresponding 3-web are autoparallel submanifolds. The torsion and curvature tensorfields satisfy some important identities, which are consequences of the defining relations of the Chern connection and of the Bianchi identities.

## 1.1 The tangent structure of a 3-web

An $n$-codimensional foliation $\mathcal{F}$ on a manifold $M$ of dimension $2n$ is a partition $\{F_\phi, \phi \in \Phi\}$ of $M$ into connected subsets such that for every point of $M$ there is an open coordinate neighbourhood $U$ with local coordinates $(x^1, ..., x^n; y^1, ..., y^n)$ in which the connected components of $U \cap F_\phi$ are determined by the equations $y^1 = const, ..., y^n = const$. We assume now that there are given two $n$-codimensional foliations $\mathcal{F} = \{F_\phi, \ \phi \in \Phi\}$ and $\mathcal{G} = \{G_\psi, \ \psi \in \Psi\}$ such that the tangent space $T_p M$ of an intersection point $p$ of the leaves $F_\phi$ and $G_\psi$ is the direct sum of the tangent spaces of these leaves: $T_p M = T_p F_\phi \oplus T_p G_\psi$. Then, in this coordinate system $(x^1, ..., x^n; y^1, ..., y^n)$ on the neighbourhood $U$, the connected components of $U \cap G_\psi$ can be given by the equations $g^1(x^1, ..., x^n; y^1, ..., y^n) = c^1, ..., g^n(x^1, ..., x^n; y^1, ..., y^n) = c^n$, where $g^1, .., g^n$ are smooth functions and $c^1, .., c^n$ are constant. Since the leaf $G_\alpha$ of the foliation $\mathcal{G}$ through the point $p$ is transversal to the leaf $F_\beta$ of the foliation $\mathcal{F}$ through $p$, it follows that the Jacobian $\det(\frac{\partial g^i}{\partial x^k}) \neq 0$ in this point $p$. We can consider the coordinate transformation $\bar{x}^i = g^i(x^1, ..., x^n; y^1, ..., y^n)$, $\bar{y}^i = y^i$, $(i = 1, ..., n)$, which gives a coordinate system $\bar{U}$ in a neighbourhood of $p$ such that the connected components of $\bar{U} \cap F_\phi$, $\phi \in \Phi$ and $\bar{U} \cap G_\psi$, $\psi \in \Psi$ are given by the equations $\bar{y}^1 = const, ..., \bar{y}^n = const$ and by $\bar{x}^1 = const, ..., \bar{x}^n = const$, respectively. If we have a third $n$-codimensional foliation $\mathcal{K} = \{K_\sigma, \sigma \in \Sigma\}$ such that the tangent space $T_p K_\sigma$ of the leaf $K_\sigma$ through $p$ is complementary to the tangent space $T_p F_\phi$ as well as to $T_p G_\psi$ then there exists a coordinate neighbourhood $W$ of $p$ in which the connected components of $W \cap F_\phi$, $\phi \in \Phi$, $W \cap G_\psi$, $\psi \in \Psi$ and of $W \cap K_\sigma$, $\sigma \in \Sigma$ are given by the equations $\bar{y}^1 = const, ..., \bar{y}^n = const$, by $\bar{x}^1 = const, ..., \bar{x}^n = const$ and by $f^i(\bar{x}^1, ..., \bar{x}^n; \bar{y}^1, ..., \bar{y}^n) = const$, respectively, where $f^1, ..., f^n$ are smooth functions. In this coordinate neighbourhood $W$ each pair of leaves of different foliations has exactly one point in common.

**Definition 1.1** A *differentiable 3-web* on a manifold $M$ is an ordered triple $(\mathcal{F}_1, \mathcal{F}_2, \mathcal{F}_3)$ of foliations such that the tangent spaces of the leaves of any two different foliations $\mathcal{F}_\alpha, \mathcal{F}_\beta (\alpha \neq \beta)$ through any point of $M$ are complementary subspaces of the tangent vector space of $M$. For a 3-web $(\mathcal{F}_1, \mathcal{F}_2, \mathcal{F}_3)$ we call the foliations $(\mathcal{F}_1, \ \mathcal{F}_2, \ \mathcal{F}_3)$ and their leaves *horizontal, vertical* and *transversal*, respectively.

**Definition 1.2** A 3-web on a manifold $M$ is called a *differentiable 3-net* if the leaves of different foliations of the 3-web intersect exactly in one point. The projection operators of the tangent bundle $TM$ onto the horizontal and vertical tangent distributions $T^{(h)} M$ and $T^{(v)} M$ with respect to the decomposition $TM = T^{(h)}M \oplus T^{(v)}M$ determine the (1,1)-tensorfields $h : TM \to T^{(h)}M$ and $v : TM \to T^{(v)}M$ on $M$ satisfying $h^2 = h$, $v^2 = v$, $h + v = id$ and $T^{(h)}M = \operatorname{Ker} v$, $T^{(v)}M = \operatorname{Ker} h$. The tensorfields $h$ and $v$ are called the horizontal and vertical projections of the tangent bundle $TM$. We denote by $j$ the (1,1)-tensorfield on $TM$ defined as follows:

if **u** is a horizontal tangent vector then $j\mathbf{u}$ is vertical such that $j\mathbf{u} - \mathbf{u}$ is transversal,

if **v** is a vertical tangent vector then $j\mathbf{v}$ is horizontal such that $j\mathbf{v} - \mathbf{v}$ is transversal,

if **w** is an arbitrary tangent vector then we set $j\mathbf{w} = jh\mathbf{w} + jv\mathbf{w}$.

The tensorfield $j$ satisfies evidently $j^2 = id$, $jh = vj$ and $T^{(t)}M = \operatorname{Ker}(j + id)$.

**Definition 1.3** An $(h, j)$-*structure* on the manifold $M$ is defined to be a pair of (1,1)-tensorfields $h$ and $j$ satisfying

$$h^2 = h, \qquad j^2 = id, \qquad hj + jh = j.$$

We write $v := id - h$ and say that $h$ and $v$ are *horizontal* and *vertical projections* on $TM$.

**Theorem 1.4** *Let $M$ be a manifold equipped with an $(h, j)$-structure such that the distributions* $\operatorname{Ker} h$, $\operatorname{Ker} v$ *and* $\operatorname{Ker}(j + id)$ *are integrable. Then the ordered triple of foliations with tangent distributions $T^{(h)}M = \operatorname{Ker} v$, $T^{(v)}M = \operatorname{Ker} h$ and $T^{(t)}M = \operatorname{Ker}(j + id)$ determine a 3-web on the manifold $M$.*

*Proof.* Since $h^2 = h$ and $v^2 = v = id - h$ the distributions $T^{(h)}M = \operatorname{Ker} v$ and $T^{(v)}M = \operatorname{Ker} h$ are complementary distributions. Moreover, we get from $hj + jh = j$ and from $vj + jv = (id - h)j + j(id - h) = j$ that the tangent bundle endomorphism $j$ induces a vector bundle isomorphism $T^{(h)}M \to T^{(v)}M$ and hence the distribution $T^{(t)}M = \operatorname{Ker}(j + id) = \{(j\mathbf{u} - \mathbf{u}) : \mathbf{u} \in T^{(h)}M\}$ is complementary to both of the distributions $T^{(h)}M = \operatorname{Ker} v$ and $T^{(v)}M = \operatorname{Ker} h$. $\square$

**Definition 1.5** If the distributions $T^{(h)}M = \operatorname{Ker} v$, $T^{(v)}M = \operatorname{Ker} h$ and $T^{(t)}M = \operatorname{Ker}(j + id)$ are integrable on a manifold equipped with an $(h, j)$-structure then the 3-web determined by the integral foliations of these distributions is called the *3-web corresponding to the $(h, j)$-structure.*

## 1.2   Chern connection

**Theorem 1.6** *For any $(h, j)$-structure on the manifold $M$ there exists a unique affine connection*

$$\nabla_X Y = h\{j[hX, jhY] + [vX, hY]\} + v\{j[vX, jvY] + [hX, vY]\} \qquad (1)$$

*satisfying*

$$\nabla h = \nabla j = 0, \quad T(hX, vY) = 0,$$

*for all vectorfields $X$ and $Y$ on $M$, where $T$ is the torsion tensorfield of the connection $\nabla$.*

*Proof.* An immediate calculation shows that the expression given in the theorem defines a covariant derivation satisfying $\nabla h = \nabla j = 0$ and $T(hX, vY) = 0$. Now we prove in three steps that a covariant derivation with these properties on a manifold equipped with a $(h, j)$-structure can be expressed by the formula (1).

1. If the vectorfield $X$ is horizontal $(vX = 0)$ and $Y$ is vertical $(hY = 0)$ then we have

$$T(X, Y) = \nabla_X Y - \nabla_Y X - [X, Y] = 0.$$

Since $0 = hY$ we get $\nabla_X hY = h\nabla_X Y = 0$, that is, $\nabla_X Y$ is vertical. Similarly, $\nabla_Y X$ is horizontal. Thus we obtain the formulae

$$\nabla_X Y = v[X, Y], \quad \nabla_Y X = h[Y, X].$$

2. If the vectorfields $X, Y$ are both horizontal $(vX = vY = 0)$, then $jY$ is vertical and the previous expressions can be applied to yield

$$\nabla_X jY = v[X, jY].$$

Thus we have in this case

$$\nabla_X Y = j\nabla_X jY = jv[X, jY] = hj[X, jY].$$

3. If the vectorfields $X, Y$ are vertical, we get similarly

$$\nabla_X Y = j\nabla_X jY = jh[X, jY] = vj[X, jY].$$

Thus the theorem is proved. $\qquad\qquad\qquad\qquad\qquad\qquad\qquad\quad \square$

**Definition 1.7** The linear connection defined by the covariant derivation $\nabla$ given by (1) is called the *Chern connection* of the $(h, j)$-structure. If the $(h, j)$-structure corresponds to a 3-web then $\nabla$ is the Chern connection of this 3-web.

A submanifold $S$ of a manifold $M$ with connection $\nabla$ is called *autoparallel*, if for any tangent vectorfields $X, Y$ tangent to $S$ the covariant derivative $\nabla_X Y$ is also a tangent vectorfield to $S$.

**Theorem 1.8** *The leaves of the foliations of a 3-web are autoparallel submanifolds with respect to the Chern connection.*

*Proof.* Since for horizontal {vertical} vectorfields $X, Y$ the covariant derivative $\nabla_X Y$ is also horizontal {vertical}, the assertion is true for horizontal and vertical leaves. For transversal vectorfields of the form $X - jX$, $Y - jY$, where $X$ and $Y$ are horizontal vectorfields, we have

$$\nabla_{X-jX}(Y - jY) = \nabla_{X-jX}Y - j\nabla_{X-jX}Y,$$

where $\nabla_{X-jX}Y$ is a horizontal vectorfield. Hence $\nabla_{X-jX}(Y - jY)$ is transversal, which proves the assertion. $\qquad\square$

### 1.3 Torsion and curvature

**Theorem 1.9** *The $(h, j)$-structure given on a manifold $M$ corresponds to a 3-web if and only if the torsion tensorfield of its Chern connection satisfies the following identities:*

$$vT(hX, hY) = 0, \quad hT(vX, vY) = 0, \quad jhT(hX, hY) = -vT(jhX, jhY)$$

*for any vectorfields $X, Y$ on $M$.*

*Proof.* Let $X, Y$ be horizontal vectorfields. It follows from $\nabla h = \nabla v = 0$ that we have $vT(X, Y) = v(\nabla_X Y - \nabla_Y X - [X, Y]) = -v[X, Y]$. Hence, according to Frobenius' Theorem, the distribution $T^{(h)}M = \text{Ker } v$ is integrable if and only if $vT(X, Y) = 0$. Similarly, $hT(X, Y) = -h[X, Y]$ for vertical vectorfields $X, Y$, which means that the vertical distribution $T^{(v)}M = \text{Ker } h$ is integrable if and only if $hT(X, Y) = 0$. Now we consider two transversal vectorfields of the form $X - jX, Y - jY$, where $X$ and $Y$ are horizontal. The distribution $\text{Ker } (j + id)$ is integrable if and only if the vectorfield $[X - jX, Y - jY]$ is transversal, too. This is satisfied if it can be written in the form $Z - jZ$ for the horizontal vectorfield $Z = h[X - jX, Y - jY]$, and hence its vertical part must have the form $v[X - jX, Y - jY] = jh[X - jX, Y - jY]$. Using the first two identities this equation can be written in the form

$$-jh\{[X, jY] - [Y, jX] - [X, Y]\} = v\{[jX, Y] - [jY, X] - [jX, jY]\},$$

or equivalently,

$$-jhT(X, Y) = vT(jX, jY),$$

and the theorem is proved. □

We can immediately derive the following identities for the curvature tensor $R(X,Y)Z = \nabla_X \nabla_Y Z - \nabla_Y \nabla_X Z - \nabla_{[X,Y]} Z$ of the Chern connection:

$$R(X,Y)h = hR(X,Y), \ R(X,Y)v = vR(X,Y), \ R(X,Y)j = jR(X,Y).$$

If the vectorfields $X, Y$ are horizontal and $Z$ is vertical, we have

$$\nabla_X \nabla_Y Z = \nabla_X (v[Y,Z]) = v[X,v[Y,Z]].$$

Since the horizontal distribution is integrable we can write $v[X,h[Y,Z]] = 0$ and hence

$$\nabla_X \nabla_Y Z = v[X,v[Y,Z]] = v[X,[Y,Z]].$$

It follows that

$$R(X,Y)Z = v\{[X,[Y,Z]] + [Y,[Z,X]] + [Z,[Y,X]]\} = 0.$$

If $X, Y, Z$ are horizontal then $jZ$ is vertical and $R(X,Y)Z = jR(X,Y)jZ = 0$. We get analogous identities for vertical vectorfields $X, Y$ similarly. Thus we obtain

**Theorem 1.10** *The curvature tensor $R$ of the Chern connection of a 3-web satisfies the identities*

$$Rh = hR, \ Rv = vR, \ Rj = jR, \ R(hX,hY) = R(vX,vY) = 0$$

*for all vectorfields $X, Y$.*

We consider now the Bianchi identities. The first of these is

$$\sigma\{R(X,Y)Z\} = \sigma\{T(T(X,Y),Z) + (\nabla_X T)(Y,Z)\},$$

where $\sigma$ denotes the summation for all cyclic permutations of the variables $X, Y, Z$.

If $X, Y, Z$ are horizontal (or vertical) we have $\sigma\{R(X,Y)Z\} = 0$, that is,

$$\sigma\{T(T(X,Y),Z) + (\nabla_X T)(Y,Z)\} = 0.$$

If $X, Y$ are horizontal (vertical) and $Z$ is vertical (horizontal) then

$$\sigma\{T(T(X,Y),Z)\} = 0, \ \sigma\{(\nabla_X T)(Y,Z)\} = (\nabla_Z T)(X,Y)$$

and $R(X,Y) = 0$. Thus we get

$$(\nabla_Z T)(X,Y) = R(Z,X)Y - R(Z,Y)X$$

and hence

$$(\nabla_Z T)(jX, jY) = -j(\nabla_Z T)(X, Y) = R(Z, Y)jX - R(Z, X)jY.$$

If $X, Y, Z$ are horizontal (vertical), then we get from this equation that

$$(\nabla_Z T)(X, Y) = R(Z, jY)X - R(Z, jX)Y.$$

Consequently we obtain $\sigma\{T(T(X, Y), Z) + R(X, jZ)Y - R(X, jY)Z\} = 0$.

The second Bianchi identity is $\sigma\{(\nabla_X R)(Y, Z) = R(T(X, Y), Z)\}$. It gives a nontrivial relation only in the case of $vX = vY = hZ = 0$ or $hX = hY = vZ = 0$. Then we get $R(T(X, Y), Z) = (\nabla_Y R)(X, Z) - (\nabla_X R)(Y, Z)$.

Thus we have proved the following

**Theorem 1.11** *The first Bianchi identity for a 3-web is equivalent to the following formulae:*

$$\sigma\{T(T(X, Y), Z) + R(X, jZ)Y - R(X, jY)Z\} = 0,$$

$$(\nabla_Z T)(X, Y) =$$

$$= R(hZ, jhY)hX - R(hZ, jhX)hY + R(vZ, hX)hY - R(vZ, hY)hX +$$

$$+ R(hZ, vX)vY - R(hZ, vY)vX + R(vZ, jvY)vX - R(vZ, jvX)vY.$$

*The second Bianchi identity is equivalent to the identity:*

$$R(T(X, Y), Z) = (\nabla_{hY} R)(hX, vZ) - (\nabla_{hX} R)(hY, vZ) +$$

$$+ (\nabla_{vY} R)(vX, hZ) - (\nabla_{vX} R)(vY, hZ),$$

*where $\sigma$ denotes the summation for all cyclic permutations of the variables $X$, $Y$ and $Z$.*

## 2  Invariant webs on Lie groups

The basic nontrivial models for the study of the global structure of 3-webs can be constructed as left invariant 3-webs on Lie groups and their factor manifolds by totally disconnected actions of groups of web-automorphisms. This chapter is devoted to the full classification of such structures on connected manifolds.

First, we introduce the covering theory of web-manifolds, using the covering theory of linear connections and the rigidity of affine maps between connected manifolds equipped with a linear connection.

The Chern connection of a left invariant web on a Lie group is without curvature. This property defines the class of group 3-webs. If the Chern connection of a group 3-web defined on a simply connected manifold has a complete Chern connection, then it is left invariant on the square manifold $G \times G$ of a Lie group $G$. If the fundamental group of such a 3-web is nontrivial then its universal covering web is defined on a square $G \times G$, and its covering transformation group is a totally disconnected group of automorphisms of the universal covering 3-web. The parallelizable 3-webs form the subclass of group 3-webs, which corresponds to abelian Lie groups.

## 2.1 Covering 3-webs

First we summarize some notions and facts about coverings of differentiable manifolds (cf. [27] Ch.1. §1.8).

Let $M$ be a connected differentiable manifold. A *differentiable cover* of $M$ consists of a connected differentiable manifold $\tilde{M}$, the *covering manifold*, and a local diffeomorphism $p : \tilde{M} \to M$, the *projection*. For each $x$ in $M$, $p^{-1}x$ is called the *fibre* over $x$. It is always a discrete space. Let $p_1 : \tilde{M}_1 \to M$ and $p_2 : \tilde{M}_2 \to M$ be two differentiable covers of $M$. Then a *covering isomorphism* from $\tilde{M}_1$ to $\tilde{M}_2$ is a diffeomorphism $\phi : \tilde{M}_1 \to \tilde{M}_2$ satisfying $p_1 = p_2 \phi$. If the covers $p_1 : \tilde{M}_1 \to M$ and $p_2 : \tilde{M}_2 \to M$ coincide then a covering isomorphism is called a *covering transformation*. Note that the set of covering transformations forms a group. The covering $p : \tilde{M} \to M$ is called a *universal covering* if the covering manifold $\tilde{M}$ is simply connected. It is well known that for any connected manifold there exists a universal covering manifold which is unique up to covering isomorphism.

A group action $\Gamma$ on a connected manifold $M$ is called *free* if it is without fixed points. The action is said to be *totally disconnected* if every point of $M$ has a neighbourhood $U$ which is disjoint from all the translates $gU$ where $g \in \Gamma$ and $g \neq 1$. The following fact is well known:

*The group of covering transformations of the universal cover of the manifold $M$ is canonically isomorphic to its fundamental group; the action of the group of covering transformations is free and totally disconnected. Conversely, if a group $\Gamma$ of diffeomorphisms of a connected and simply connected manifold $\tilde{M}$ acts freely and in a totally disconnected way, then the map $\tilde{M} \to \tilde{M}/\Gamma$ from $\tilde{M}$ to the manifold $M = \tilde{M}/\Gamma$ of orbits is a differentiable covering and the fundamental group of $M$ is isomorphic to $\Gamma$.*

Let $M$ be a manifold equipped with a connection $\nabla$. Since any cover $\tilde{M}$ is locally diffeomorphic to $M$, the connection $\nabla$ can be lifted to a unique connection $\tilde{\nabla}$ on $\tilde{M}$ such that the projection $p : \tilde{M} \to M$ commutes with the covariant derivation, that is $p$ is an affine map. From the construction of the connection $\tilde{\nabla}$ on the cover $\tilde{M}$ follows immediately that the covering isomorphisms between covers are affine maps with respect to the lifted connections. In particular, if $\tilde{M}$ is the universal cover of $M$ then the group of covering maps is an affine transformation group on $\tilde{M}$ with respect to the lifted connection $\tilde{\nabla}$ on $\tilde{M}$, such that its action is free and totally disconnected. Conversely, if there is given an affine transformation group $G$ on a connected and simply connected manifold $\tilde{M}$ with connection $\tilde{\nabla}$ with free and totally disconnected action, then the covering map $p : \tilde{M} \to \tilde{M}/G$ is an affine map onto the manifold $M = \tilde{M}/G$ of orbits with respect to the connection $\nabla$ induced by $\tilde{\nabla}$ on the factor manifold $M$.

Now let $M$ and $\bar{M}$ be connected differentiable 3-web manifolds equipped with the corresponding $(h, j)$ and $(\bar{h}, \bar{j})$-structures, respectively. Let $\nabla$ and $\bar{\nabla}$ denote the Chern connections of $M$ and $\bar{M}$, respectively. A differentiable map $\phi : M \to \bar{M}$ of the 3-web on $M$ to the 3-web on $\bar{M}$ is called *foliation preserving* if $\phi$ maps the leaves of the foliations of $M$ into the leaves of the foliations on $\bar{M}$. If $M = \bar{M}$, then a foliation preserving diffeomorphism $\phi : M \to M$ is called a *foliation preserving automorphism*. A differentiable map $\phi : M \to \bar{M}$ is called a *web-homomorphism* if $\phi$ maps the horizontal, vertical or transversal leaves of the foliations of $M$ into the horizontal, vertical or transversal leaves, respectively, of the manifold $\bar{M}$. If a web-homomorphism is a diffeomorphism, it is called a *web-isomorphism* and if $M = \bar{M}$ then it is called a *web-automorphism*.

**Proposition 2.1** *Let $M$ and $M'$ be 3-web manifolds equipped with the associated $(h, j)$ and $(h', j')$-structures. If the differentiable map $\psi : M \to M'$ is a local diffeomorphism then it is a web-homomorphism if and only if the following conditions are satisfied:*

1. *there exists a point $x \in M$ such that the map $(\psi_*)_x : T_x M \to T_{\psi(x)} M'$ is a linear isomorphism between the $(h_x, j_x)$-structure on $T_x M$ and the $(h'_{\psi(x)}, j'_{\psi(x)})$-structure on $T_{\psi(x)} M'$,*

2. *$\psi$ is an affine map with respect to the Chern connections associated with the $(h, j)$-structures of the manifolds $M$ and $M'$.*

*Proof.* Clearly, a web-homomorphism is a linear homomorphism from the $(h_z, j_z)$-structure on $T_z M$ to the $(h'_{\psi(z)}, j'_{\psi(z)})$-structure on $T_{\psi(z)} M'$ for any $z \in M$ and hence it is an affine map with respect to the corresponding Chern connections. Conversely, let $\psi : M \to M'$ be an affine map with respect to the Chern connections of $M$ and $M'$, and let us assume that at the point $x \in M$ the map $(\psi_*)_x : T_x M \to T_{\psi(x)} M'$ is a homomorphism between the $(h_x, j_x)$ and the $(h'_{\psi(x)}, j'_{\psi(x)})$-structures. Given $y \in M$, choose a smooth path $\sigma$ in $M$

from $x$ to $y$ and let $\sigma' = \psi\sigma$. As the map $\psi$ is affine, the tangent map $(\psi_*)_y$ : $T_yM \to T_{\psi(y)}M'$ can be written in the form $(\psi_*)_y = \tau'(\psi_*)_x\tau^{-1}$, where $\tau$ and $\tau'$ denote the parallel translation along the paths $\sigma$ and $\sigma'$, respectively. Since the tensorfields $h, h'$ and $j, j'$ satisfy $\nabla h = \nabla h' = \nabla j = \nabla j' = 0$ the parallel translations $\tau$ or $\tau'$ are linear isomorphisms between the $(h_x, j_x)$ and the $(h_y, j_y)$-structures or the $(h'_{\psi(x)}, j'_{\psi(x)})$ and the $(h'_{\psi(y)}, j'_{\psi(y)})$-structures of the tangent spaces $T_xM$ and $T_yM$ or $T_{\psi(x)}M'$ and $T_{\psi(y)}M'$, respectively. Hence the affine map $\psi : M \to M'$ is a linear homomorphism from the $(h_y, j_y)$-structure on $T_yM$ to $(h'_{\psi(y)}, j'_{\psi(y)})$-structure on $T_{\psi(y)}M'$ for any $y \in M$. It follows that $\psi$ is a web-homomorphism. $\square$

**Theorem 2.2** *Let $M$ be a connected manifold equipped with a 3-web structure. The universal covering manifold $M'$ of $M$ has a uniquely determined 3-web structure, for which the projection $p : M' \to M$ is a web-homomorphism. The group of covering transformations of the covering $p : M' \to M$ consists of web-automorphisms. Conversely, if there is given a group $\Gamma$ of web-automorphisms on a connected and simply connected 3-web manifold $M'$ with a free and totally disconnected action, then the covering map $p : M' \to M'/\Gamma$ is a web-homomorphism onto the manifold $M = M'/\Gamma$ of orbits with respect to the 3-web structure induced on the factor manifold $M$.*

*Proof.* Let us choose the horizontal, vertical and transversal distributions on $M$ and consider the corresponding $(h, j)$-structure on $M$. Since the map $p : M' \to M$ is a local diffeomorphism, we can define a $(h', j')$-structure on the universal covering manifold $M'$ satisfying $p_*h' = hp_*$ and $p_*j' = jp_*$. The distributions corresponding to the $(h', j')$-structure on $M'$ are integrable, since the mapping $p_*$ maps them onto the tangent distributions of the foliations of the 3-web on $M$. Consequently, the $(h', j')$-structure on $M'$ determines a 3-web on $M'$, and the projection mapping $p$ maps the leaves of the foliations on $M'$ onto the leaves of the corresponding foliations on $M$ and hence $p$ is a web-homomorphism. The covering transformations induce the identity map on $M$, from which follows that they are web-automorphisms of the $(h', j')$-structure, and hence of the web structure of $M'$.

Conversely, if there is given a free and totally disconnected action of the group $\Gamma$ consisting of web-automorphisms of the 3-web on $M'$ corresponding to the $(h', j')$-structure, then we can define a $(h, j)$-structure on the factor manifold $M = M'/G$ such that the projection mapping $p : M' \to M$ satisfies $p_*h' = hp_*$ and $p_*j' = jp_*$, and hence it is a web-homomorphism. $\square$

## 2.2 Group 3-webs

We recall that a connection is called complete if every maximal geodesic has the form $\{\gamma(t), -\infty < t < +\infty\}$ where $t$ is the affine parameter of $\gamma$.

**Proposition 2.3** *Let $\nabla$ be the Chern connection of a 3-web on the connected and simply connected manifold $M$ and assume that its curvature tensorfield vanishes. If the connection $\nabla$ is complete then $M$ is diffeomorphic to a Lie group $K$ such that the tangent distributions of the foliations of the 3-web on $K$, which is induced by the diffeomorphism $M \to K$, are left invariant. The Chern connection of the 3-web on the Lie group $K$ coincides with its left invariant connection.*

*Proof.* The first Bianchi identity (cf. Theorem 1.10) implies $\nabla T = 0$. It follows from the theory of invariant connections on reductive homogeneous spaces ($^{15}$ Chapter X. Theorem 2.8 and Proposition 2.12) that if we choose an origin $e \in M$, then we can identify the manifold $M$ with a simply connected Lie group $K$, acting sharply transitively on $M$ and the origin $e$ with the unit of $K$. The Lie algebra operation $[X, Y]$ on the tangent space $\mathfrak{K} = T_e K$ has the form $[X, Y] = -T(X, Y)$. The Chern connection $\nabla$ will be identified with the left invariant connection of $K$. Hence the parallel translation on $K$ coincides with the tangent map of the left multiplication. Since the horizontal, vertical and transversal distributions are invariant with respect to the parallel translation we obtain that they are left invariant. $\square$

**Proposition 2.4** *A. The Lie algebra on the tangent space $\mathfrak{K} = T_e K$ of the Lie group $K$, acting sharply transitively on $M$, decomposes into the direct sum $\mathfrak{K} = \mathfrak{K}^{(h)} \oplus \mathfrak{K}^{(v)}$ of Lie subalgebras induced on the horizontal and vertical subspaces of $\mathfrak{K} = T_e K$. The restriction $\theta : \mathfrak{K}^{(h)} \to \mathfrak{K}^{(v)}$ of the map $-j_e : \mathfrak{K} \to \mathfrak{K}$ is a Lie algebra isomorphism of the Lie subalgebra $\mathfrak{K}^{(h)}$ onto $\mathfrak{K}^{(v)}$.*

*B. The Lie group $K$, acting sharply transitively on $M$, decomposes into the direct product $K = K^{(h)} \times K^{(v)}$ of the Lie subgroups $K^{(h)} = \exp \mathfrak{K}^{(h)}$ and $K^{(v)} = \exp \mathfrak{K}^{(v)}$. The map $\Theta = \exp \cdot \theta \cdot \exp^{-1} : K^{(h)} \to K^{(v)}$ is a Lie group isomorphism of the Lie subgroup $K^{(h)}$ onto $K^{(v)}$.*

*Proof.* The assertion A. follows from $T(hX, vY) = 0$ and from the identities in Theorem 1.9. Namely if $\mathbf{x}, \mathbf{y} \in \mathfrak{K}$ then we have

$$[\theta\mathbf{x}, \theta\mathbf{y}] = [-j_e\mathbf{x}, -j_e\mathbf{y}] = -T(j_e\mathbf{x}, j_e\mathbf{y}) = j_e T(\mathbf{x}, \mathbf{y}) = \theta[\mathbf{x}, \mathbf{y}].$$

Since the groups $K, K^{(h)}$ and $K^{(v)}$ are simply connected the direct sum decomposition of the Lie algebra $\mathfrak{K}$ implies the direct product decomposition of the Lie group $K$. $\square$

**Theorem 2.5** *For a 3-web with complete Chern connection on a connected and simply connected manifold M there exists a simply connected Lie group G and a 3-net structure on the direct product manifold $G \times G$ with horizontal, vertical and transversal foliations given by the submanifolds $G \times \{g\}$, $\{g\} \times G$, $g \in G$ and by the left cosets of the diagonal subgroup $\Delta G = \{(g, g) : g \in G\}$ respectively, which is isomorphic to the 3-web on M, if and only if the curvature tensorfield of the Chern connection vanishes.*

*Proof.* It follows from the previous propositions that if the curvature of the Chern connection vanishes then the manifold $M$ can be identified with the Lie group $K = K^{(h)} \times K^{(v)}$, where $K^{(h)}$ and $K^{(v)}$ are the simply connected Lie subgroups corresponding to the Lie subalgebras $\mathfrak{K}^{(h)}$ and $\mathfrak{K}^{(v)}$, respectively. Moreover the map $\Theta : K^{(h)} \to K^{(v)}$ is a group isomorphism. Now we put $G = K^{(h)}$ and consider the 3-web on $G \times G$, which is the preimage of the 3-web on $K = K^{(h)} \times K^{(v)}$. Since the tangent distributions of the 3-web on $K = K^{(h)} \times K^{(v)}$ are left invariant and $T_e^{(h)} K = \mathfrak{K}^{(h)}$, $T_e^{(v)} K = \mathfrak{K}^{(v)}$, the horizontal and vertical leaves of the 3-web on $G \times G$ can be represented in the form $G \times \{g\}$ and $\{g\} \times G$, $g \in G$, respectively. The linear map $j'_{(e,e)} = ([(id, \Theta)_*^{-1} j(id, \Theta)]_{(e,e)})_*$ on $T_{(e,e)}(G \times G)$ corresponds to the map $j_e$ on $T_{(e)} K$, and it acts on the tangent vectors $(\mathbf{x}, \mathbf{y}) \in T_{(e,e)}(G \times G)$ at the unit $(e, e) \in G \times G$ in the following way:

$$j'_{(e,e)}(\mathbf{x}, \mathbf{y}) = [(id, \Theta)_*^{-1} j(id, \Theta)_*]_{(e,e)}(\mathbf{x}, \mathbf{y}) = [(id, \theta) j(id, \theta)]_{(e,e)}(\mathbf{x}, \mathbf{y}) =$$

$$= [(id, \theta) j]_{(e,e)}(\mathbf{x}, -j_e \mathbf{y}) = [(id, \theta)]_{(e,e)}(-\mathbf{y}, j_e \mathbf{x}, ) = (-\mathbf{y}, -\mathbf{x}, )$$

since $\Theta_*(\mathbf{x}) = \theta(\mathbf{x}) = -j_e(\mathbf{x})$ and $j_{(e,e)}(\mathbf{u}, \mathbf{v}) = (j_e \mathbf{v}, j_e \mathbf{u})$ for any $\mathbf{x}, \mathbf{u}, \mathbf{v} \in T_e G$. Hence the transversal subspace at $(e, e) \in G \times G$ has the form

$$T_{(e,e)}^{(t)}(G \times G) = \{(\mathbf{x}, 0) - j'_{(e,e)}(\mathbf{x}, 0) = (\mathbf{x}, \mathbf{x}), \ \mathbf{x} \in T_e G\}.$$

The transversal distribution is left invariant on the group $G \times G$ and hence it can be written in the form

$$T_{(x,y)}^{(t)}(G \times G) = \{((\lambda_x)_* \mathbf{x}, (\lambda_y)_* \mathbf{x}), \ \mathbf{x} \in T_e G, \ x, y \in G\},$$

where $\lambda_x$ denotes the left multiplication map in the group $G$. It follows that the transversal leaf through $(e, e)$ is the diagonal subgroup $\Delta G = \{(g, g) \in G \times G : g \in G\}$ since its tangent spaces have the previous form and the left cosets of this subgroup are the transversal leaves of $G \times G$.

Conversely, for any group $G$, the leaves of different foliations on the 3-web determined by the above construction on $G \times G$ have a unique intersection point and hence they define a 3-net on $G \times G$. Since the horizontal, vertical and transversal distributions are left invariant on the group $G \times G$, the Chern connection of this 3-net is also left invariant and hence its curvature vanishes. $\square$

Considering the diffeomorphism $(x,y) \mapsto (x,y^{-1}) : G \times G \to G \times G$ we obtain the following equivalent representation of simply connected 3-webs with vanishing curvature tensorfield:

**Theorem 2.6** *Any connected and simply connected 3-web with complete Chern connection and vanishing curvature tensorfield is isomorphic to the 3-net on the manifold $G \times G$ for a Lie group $G$, the horizontal, vertical and transversal leaves of which are given by the submanifolds: $G \times \{g\}$, $\{g\} \times G$, $g \in G$ and by the level sets $\{(x,y) : xy = const, x,y \in G\}$ of the multiplication function $(x,y) \mapsto xy : G \times G \to G$, respectively.*

*Proof.* The diffeomorphism given by $(x,y) \mapsto (x,y^{-1})$ maps the left cosets

$$(x,y)\Delta G = \{(xg,yg) \in G \times G : \ g \in G\},$$

for $x,y \in G$ onto the sets

$$\{(xg,(yg)^{-1}) \in G \times G : \ g \in G\} = \{(u,v) : \ uv = const\},$$

where the *const* is given by $const = xy^{-1}$. $\square$

**Definition 2.7** The 3-net given by the horizontal, vertical and transversal foliations $G \times \{g\}$, $\{g\} \times G$, $g \in G$ and by the level sets $\{(x,y) : xy = const, x,y \in G\}$ of the multiplication function $(x,y) \mapsto xy : G \times G \to G$, respectively, is called the *group 3-net determined by the Lie group $G$* on the manifold $G \times G$.

Applying the results of Theorem 2.6 to the universal covering 3-webs (cf. Theorem 2.2) we get the following

**Theorem 2.8** *Any connected 3-web with complete Chern connection and vanishing curvature tensorfield is isomorphic to the 3-web on the quotient manifold $(G \times G)/\Gamma$ of a connected and simply connected group 3-net determined by the Lie group $G$ with respect to the free and totally disconnected action of a group $\Gamma$ of direction preserving web-automorphisms.*

We recall that the group of web-automorphisms of a 3-net determined by a group $G$ can be described in the following way:

**Proposition 2.9** *The group of web-automorphisms of a 3-net on the manifold $G \times G$ for a Lie group $G$ determined as in the previous theorem is isomorphic to the semi-direct product of the group $G \times G$ with the group $Aut(G)$ of automorphisms of $G$ acting on $G \times G$ by $\alpha : (x,y) \mapsto (\alpha(x), \alpha(y))$ for any $x, y \in G$ and $\alpha \in Aut(G)$.*

*Proof.* A mapping $\psi : G \times G \to G \times G$ can be written in the form $(x,y) \mapsto (\beta(x,y), \gamma(x,y))$. If $\psi$ is a web-automorphism, it preserves the horizontal and vertical foliations and hence it has the form $(x,y) \mapsto (\beta(x), \gamma(y))$. The points $(x,y)$ of a transversal leaf satisfy $xy^{-1} = z$ or $x = zy$ for a given constant $z \in G$, hence their images $(\beta(x), \gamma(y)) = (\beta(zy), \gamma(y))$ satisfy $(\beta(zy) = \delta(z)\gamma(y)$ for a constant $\delta(z)$ depending on $z$. It follows $\beta(e) = \delta(e)\gamma(e)$ for the unit $e \in G$ and $\beta(y) = \delta(e)\gamma(y)$, $\beta(z) = \delta(z)\gamma(e)$. Consequently we have $\delta(e)\gamma(zy) = \delta(z)\gamma(y) = \beta(z)\gamma(e)^{-1}\gamma(y) = \delta(e)\gamma(z)\gamma(e)^{-1}\gamma(y)$ and hence $\gamma(e)^{-1}\gamma(zy) = \gamma(e)^{-1}\gamma(z)\gamma(e)^{-1}\gamma(y)$, which means that the map $\alpha : x \mapsto \gamma(e)^{-1}\gamma(x) : G \to G$ is an automorphism. We have $\gamma(x) = \gamma(e)\alpha(x)$ and $\beta(y) = \delta(e)\gamma(y) = \beta(e)\alpha(y)$. Clearly, for any $\beta(e), \gamma(e) \in G$ and an arbitrary automorphism $\alpha \in Aut(G)$ these expressions give a web-automorphism of $G \times G$. $\qquad\square$

From the representation of 3-webs with complete Chern connection and vanishing curvature we obtain a local representation theorem of 3-webs with vanishing curvature tensorfield without assuming the completeness of the Chern connection.

**Theorem 2.10** *The curvature tensorfield of the Chern connection of a 3-web $M$ vanishes if and only if every point of $M$ has a neighbourhood such that the induced 3-web on this neighbourhood is web-isomorphic to the induced 3-web on an open domain in a group 3-net determined by a Lie group.*

*Proof.* The assertion follows from the previous theorems and from the local affine equivalence theorem of manifolds with affine connections, for which the covariant derivatives of the torsion and curvature tensorfields vanish (cf. [14] Ch.VI. Theorem 74.). $\qquad\square$

The above results motivate the following

**Definition 2.11** A 3-web $M$ is called a *group 3-web* if every point of $M$ has a neighbourhood such that the induced 3-web on this neighbourhood is web-isomorphic to the induced 3-web on an open domain in a group 3-net determined by a Lie group.

Now we can formulate the following

**Corollary 2.12** *The curvature tensorfield of the Chern connection of a 3-web $M$ vanishes if and only if $M$ is a group 3-web.*

*2.3  Parallelizable 3-webs*

**Definition 2.13** A 3-web $M$ is called *affine,* if there exists a web-isomorphism of $M$ to the 3-web on an open domain of the $2n$-dimensional affine space determined by 3 families of parallel affine subspaces of dimension $n$. A 3-web $M$ is called *parallelizable,* if any point of $M$ has a neighbourhood such that the 3-web induced by $M$ on this neighbourhood is web-isomorphic to an affine 3-web.

Since the Chern connection of a 3-web is defined locally, the Chern connection of parallelizable 3-webs is locally affine isomorphic to the canonical flat connection of the affine space, i.e. its curvature and torsion tensorfields vanish. Consequently, the complete parallelizable 3-webs are special cases of group 3-webs and we can apply the previous theorems to their investigation.

**Theorem 2.14** *A 3-web is parallelizable if and only if its Chern connection has vanishing torsion and curvature tensorfields. The connected and simply connected parallelizable 3-webs with complete Chern connection are the affine 3-nets. The parallelizable 3-webs with complete Chern connection are just the quotients $N/\Gamma$ of an affine 3-net $N$, where $\Gamma$ is a group of web-automorphisms of $N$ acting freely and in a totally disconnected way on the affine space $N$.*

*Proof.* From the previous theorems follows that a 3-web is parallelizable if and only if it is locally isomorphic to the group 3-web determined by the vector group $G = \mathbb{R}^n$. But the group 3-web of the commutative group $G = \mathbb{R}^n$ has vanishing torsion and curvature and hence the assertion follows.  □

## 3  Closure conditions and curvature

Since any 1-codimensional distribution on a 2-manifold is integrable, 3-webs on a 2-manifold have interesting particular properties. In this case the curvature tensorfield has only one scalar invariant, which can be geometrically interpreted as a measure of the non-closeness of the classical hexagonal configurations. Indeed, according to a theorem of Blaschke, the vanishing of the curvature gives a characterization of locally hexagonal 3-webs on 2-dimensional manifolds. This case gives an important motivation for the following study of higher dimensional 3-webs. The classical local closure conditions of the configurations of Reidemeister, Bol and Moufang and of the hexagonal configuration imply identities for the curvature tensorfield of the Chern connection. In this chapter we express the measure of non-closeness of the classical configurations with help of the components of the curvature tensorfield.

For the study of the local structure of a 3-web we introduce a distinguished local coordinate system, in which the horizontal and vertical leaves are given by parallel affine subspaces and the transversal leaves are the level sets of a smooth mapping. Moreover we can assume that this mapping satisfies further conditions in a fixed origin. We compute the curvature of the Chern connection at the origin expressed by this mapping and its derivatives and describe, how the curvature tensorfield can be considered as the infinitesimal measure of the non-closeness of the Reidemeister configurations. According to our previous theory of group 3-webs, the curvature tensorfield vanishes if and only if the 3-web satisfies the Reidemeister closure condition. Since Bol configurations are particular cases of Reidemeister configurations, we can obtain from the previous results infinitesimal measures for the non-closeness of Bol configurations, there resulting necessary conditions for Bol 3-webs. Moufang webs are defined by the closeness of Bol configurations with respect to two different foliations, hence we obtain necessary curvature conditions for Moufang webs too.

The construction of distinguished coordinate systems can be translated to the language of non-associative algebra: the closure conditions correspond to weak associative identities and the torsion and curvature of the Chern connection gives infinitesimal measure of the commutativity and the associativity of the associated multiplications.

## 3.1 Distinguished coordinate systems

Let $M$ be a 3-web manifold of dimension $2n$ equipped with a $(h, j)$-structure, which is associated with the 3-web structure of $M$. Let be given a point $p_0$ of the manifold $M$. We consider a local coordinate system $(\mathbf{x}, \mathbf{y}) = (x^1, ..., x^n; y^1, ..., y^n)$ in a neighbourhood $U$ of $p_0$, such that the point $p_0$ is given by $(\mathbf{0}, \mathbf{0})$ and the connected components of the horizontal, vertical or transversal leaves have the equations $\mathbf{y} = const$, $\mathbf{x} = const$ or $\mathbf{f}(\mathbf{x}, \mathbf{y}) = const$, respectively, for a smooth mapping $\mathbf{f}$ having the coordinate functions $f^1(x^1, ..., x^n; y^1, ..., y^n) = f^1(\mathbf{x}, \mathbf{y}), ..., f^n(x^1, ..., x^n; y^1, ..., y^n) = f^n(\mathbf{x}, \mathbf{y})$. We can assume that the mapping $\mathbf{f}$ satisfies $\mathbf{f}(\mathbf{0}, \mathbf{0}) = \mathbf{0}$. Since the horizontal, vertical and transversal tangent spaces are pairwise complementary subspaces, the partial tangent operators $\mathbf{f}'_\mathbf{x}(\mathbf{0}, \mathbf{0})$ and $\mathbf{f}'_\mathbf{y}(\mathbf{0}, \mathbf{0})$ of the mapping $\mathbf{f}$ with respect to the variables $\mathbf{x}$ and $\mathbf{y}$, respectively, have maximal rank. We consider now a neighbourhood $W \subset U$ of the point $p_0$ in which the mappings defined by $\chi : \mathbf{x} \mapsto \mathbf{f}(\mathbf{x}, \mathbf{0})$ and $\omega : \mathbf{y} \mapsto \mathbf{f}(\mathbf{0}, \mathbf{y})$ are bijective.

The coordinate mapping $\chi^{-1} \cdot \omega$ determines a well-defined differentiable map $\Phi : W^v \to W^h$, where $W^v$ and $W^h$ denote the intersection of the neigh-

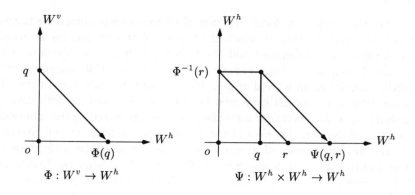

Figure 1.

bourhood $W$ with the vertical and horizontal leaf through $p_0$, respectively. Clearly, $\Phi : W^v \to W^h$ maps a point of the vertical leaf to the point of the horizontal leaf, which belongs to the same vertical leaf.

We introduce the new coordinate maps $\bar{\mathbf{x}}, \bar{\mathbf{y}}$ on $W$ by $\bar{\mathbf{x}} = \mathbf{x}, \bar{\mathbf{y}} = \chi^{-1} \cdot \omega(\mathbf{y})$ and the new mappings $\bar{\mathbf{f}}(\bar{\mathbf{x}}, \bar{\mathbf{y}})$ by $\bar{\mathbf{f}}(\bar{\mathbf{x}}, \bar{\mathbf{y}}) = \chi^{-1} \cdot \mathbf{f}(\bar{\mathbf{x}}, \omega^{-1} \cdot \chi(\bar{\mathbf{y}}))$. In this co-ordinate neighbourhood $W$ the point $p_0$ is given by $(\mathbf{0}, \mathbf{0})$ and the connected components of the horizontal, vertical or transversal leaves have the equations $\bar{\mathbf{y}} = const$, $\bar{\mathbf{x}} = const$ or $\bar{\mathbf{f}}(\bar{\mathbf{x}}, \bar{\mathbf{y}}) = const$, respectively. The coordinates $\bar{\mathbf{x}}$, $\bar{\mathbf{y}}$ and the mapping $\bar{\mathbf{f}}$ satisfy

$$\bar{\mathbf{f}}(\bar{\mathbf{x}}, \mathbf{0}) = \chi^{-1} \cdot \mathbf{f}(\bar{\mathbf{x}}, \mathbf{0}) = \bar{\mathbf{x}}, \qquad \bar{\mathbf{f}}(\mathbf{0}, \bar{\mathbf{y}}) = \chi^{-1} \cdot \mathbf{f}(\mathbf{0}, \omega^{-1} \cdot \chi(\bar{\mathbf{y}})) = \bar{\mathbf{y}}$$

and the points $(\bar{\mathbf{x}}, \mathbf{0})$, $(\mathbf{0}, \bar{\mathbf{y}})$ are contained in the same transversal leaf if and only if

$$\bar{\mathbf{x}} = \bar{\mathbf{f}}(\bar{\mathbf{x}}, \mathbf{0}) = \bar{\mathbf{f}}(\mathbf{0}, \bar{\mathbf{y}}) = \bar{\mathbf{y}}.$$

We notice that the coordinate mapping $\bar{\mathbf{f}}(\bar{\mathbf{x}}, \bar{\mathbf{y}}) = \chi^{-1} \cdot \mathbf{f}(\bar{\mathbf{x}}, \omega^{-1} \cdot \chi(\bar{\mathbf{y}}))$ deter-mines a differentiable map $\Psi : W^h \times W^h \to W^h$ such that for any $q$, $r \in W^h$ the point $\Psi(q, r)$ belongs to the same transversal leaf as the intersection point of the vertical leaf through $q$ and of the horizontal leaf through $\Phi^{-1}(r)$.

**Definition 3.1** Let $p_0$ be a given point of the 3-web manifold $M$ and let $(\mathbf{x}, \mathbf{y}) = (x^1, ..., x^n; y^1, ..., y^n)$ be a local coordinate system in a neighbour-hood $U$ of $p_0$, such that the point $p_0$ is given by $(\mathbf{0}, \mathbf{0})$ and the connected components of the horizontal, vertical or transversal leaves have the equations $\mathbf{y} = const$, $\mathbf{x} = const$ or $\mathbf{f}(\mathbf{x}, \mathbf{y}) = const$, respectively. This local coordinate system is called *a distinguished coordinate system centered at the point $p_0$*, if the coordinates **x**, **y** and the mapping **f** satisfy

$$\mathbf{f}(\mathbf{x}, 0) = \mathbf{x}, \qquad \mathbf{f}(0, \mathbf{y}) = \mathbf{y}$$

and the points $(\mathbf{x}, 0)$, $(0, \mathbf{y})$ belong to the same transversal leaf if and only if

$$\mathbf{x} = \mathbf{f}(\mathbf{x}, 0) = \mathbf{f}(0, \mathbf{y}) = \mathbf{y}.$$

We obtain from the previous construction the following

**Proposition 3.2** *For any point $p_0$ of a 3-web manifold $M$ there exists a distinguished coordinate system centered at the point $p_0$ with coordinates **x**, **y** and mapping **f**, satisfying*

$$\mathbf{f}(0, 0) = 0, \ \mathbf{f}'_{\mathbf{x}}(0, 0) = \mathbf{f}'_{\mathbf{y}}(0, 0) = id,$$

$$\mathbf{f}''_{\mathbf{xx}}(0, 0) = \mathbf{f}''_{\mathbf{yy}}(0, 0) = \mathbf{f}^{(r)}_{\mathbf{xx}\ldots\mathbf{x}}(0, 0) = \mathbf{f}^{(r)}_{\mathbf{yy}\ldots\mathbf{y}}(0, 0) = 0.$$

The construction of the distinguished coordinate system implies the following

**Theorem 3.3** *For any point $p_0$ of a 3-web manifold $M$ there exists a neighbourhood $W$ of $p_0$ which is diffeomorphic to the square $W^h \times W^h$, where $W^h$ is a neighbourhood of $p_0$ on the horizontal leaf $L$ trough $p_0$, such that the horizontal and vertical foliations induce on $W^h \times W^h$ the foliations given by the equations $\pi_2 = 0$ and $\pi_1 = 0$, respectively, where $\pi_1$ and $\pi_2$ are the projections of the product $W^h \times W^h$ onto the first or the second factor, respectively, and the transversal leaves are given by the equations $\Psi(q, r) = const$, $q, r \in W^h$, where $\Psi : W^h \times W^h \to L$ satisfies $\Psi(q, p_0) = q$, $\Psi(p_0, r) = r$.*

*Proof.* In a distinguished coordinate system centered at $p_0$ the mapping

$$(\mathbf{x}, \mathbf{y}) \mapsto ((\mathbf{x}, 0), (0, \mathbf{y}))$$

determines a local diffeomorphism $W \to W^h \times W^h$ and the function $\Psi :$ $W^h \times W^h \to L$ is given by the coordinate mapping **f** satisfying the properties described in the previous proposition. $\qquad \square$

### 3.2 Differentiable loops associated with a 3-web

**Definition 3.4** A set $L$ with a distinguished element $1 \in L$ equipped with the binary operations $\circ, /, \backslash : L \times L \to L$ satisfying the identities:

$$x \circ 1 = 1 \circ x = x, \quad x \backslash (x \circ y) = x \circ (x \backslash y) = y, \quad (x \circ y)/y = (x/y) \circ y = x$$

is called a *loop*. The element 1 is the *identity* of $L$, the operations $\circ, \backslash$ and $/$ are the *multiplication*, the *left division* and the *right division*, respectively, of the loop $L$.

If $L$ is a differentiable manifold and the operations $\circ, /, \backslash : L \times L \to L$ are differentiable maps then $L$ is called a *differentiable loop*.

If the differentiable partial operations $\circ, /, \backslash : W \times W \to L$ are defined only on a neighbourhood $W$ of $1 \in L$ and the loop identities are satisfied if the expressions on the left and on the right side of the equations are defined then $L$ is called a *differentiable local loop*.

For any differentiable loop $L$ with identity element $1 \in L$ and with operations $\circ, /, \backslash : L \times L \to L$ one can define a 3-net on the manifold $L \times L$ given by the horizontal, vertical and transversal foliations : $L \times \{l\}$, $\{l\} \times L$, $l \in L$ and by the level sets $\{(x,y) : x \circ y = const, x, y \in L\}$ of the multiplication function $(x,y) \mapsto x \circ y : L \times L \to L$, respectively. This 3-net on the manifold $L \times L$ is called *the 3-net determined by the differentiable loop $L$*.

If the differentiable partial operations $\circ, /, \backslash : W \times W \to L$ define a local loop on a neighbourhood $W$ of $1 \in L$ then we have a 3-web on the manifold $W \times W$ defined by the horizontal, vertical and transversal foliations : $W \times \{l\}$, $\{l\} \times W$, $l \in W$ and by the level sets $\{(x,y) : x \circ y = const, x, y \in W\}$ of the multiplication function $(x,y) \mapsto x \circ y : W \times W \to L$, which is called *the 3-web determined by the local loop $L$*.

**Theorem 3.5** *For any point $p_0$ of a 3-web manifold $M$ there exists a neighbourhood $W$ of $p_0$ and a differentiable local loop with identity $1$ defined on the neighbourhood $W^h$ of $1$ on the horizontal leaf $L$ through $p_0$ such that the 3-web induced on $U$ is web-isomorphic to the 3-web determined by the local loop $L$.*

*Proof.* The assertion follows from Theorem 3.3, since the coordinate function $\mathbf{f}$ of the mapping $\Psi : W^h \times W^h \to L$ satisfies

$$\mathbf{f}(0,0) = 0, \ \mathbf{f}'_{\mathbf{x}}(0,0) = \mathbf{f}'_{\mathbf{y}}(0,0) = id.$$

Hence for the local multiplication determined by $q \circ r = \Psi(q,r)$ we can define local left and right divisions according to the Implicit Function Theorem. $\square$

**Definition 3.6** The local loop described in the previous theorem, given on a neighbourhood $W^h$ of a horizontal leaf $L$ trough a point $p_0$ of a web manifold $M$ is called *the local loop centered at the point $p_0$ associated with the 3-web $M$*.

*Remark.* According to Theorem 2.10 the local loops associated with group 3-webs are local groups.

## 3.3  3-webs on 2-manifolds

Let $M$ be a 2-dimensional manifold equipped with an $(h,j)$-structure. In this case the horizontal, vertical and transversal distributions are one-dimensional and hence they are integrable, that is the $(h,j)$-structure is associated with a 3-web structure. Since the leaves of the foliations of a 3-web are totally geodesic submanifolds, the leaves of the foliations of the 3-web on $M$ are geodesic curves of the Chern connection. The torsion tensorfield is skew-symmetric and leaves invariant the horizontal and vertical distributions, hence it vanishes identically. Since the horizontal and vertical vectors are eigen-vectors of the curvature operator $R(X,Y)$, we denote by $r^{(h)}(X,Y)$ and by $r^{(v)}(X,Y)$ the corresponding eigen-values, which are skew-symmetric bilinear forms on the tangent bundle $TM$. Then the curvature tensorfield can be written in the form

$$R(X,Y)Z = R(X,Y)hZ + R(X,Y)vZ = R(X,Y)hZ + jR(X,Y)jvZ =$$

$$= r^{(h)}(X,Y)hZ + jr^{(h)}(X,Y)jvZ = r^{(h)}(X,Y)Z.$$

Similarly we get $R(X,Y)Z = r^{(v)}(X,Y)Z$ and hence $r(X,Y) = r^{(h)}(X,Y) = r^{(v)}(X,Y)$. Let $U$ and $V$ denote a horizontal and a vertical vectorfield, respectively, on the 2-manifold $M$. Then we can write
$$r(x_1U + x_2V, y_1U + y_2V) = x_1y_2r(U,V) + x_2y_1r(V,U) = (x_1y_2 - x_2y_1)r(U,V),$$
which means that the curvature is determined by the scalar component $r(U,V)$.

**Definition 3.7** Let $M$ be a 3-web manifold. A set of points of the manifold $M$ is called *(horizontally, vertically or transversally) collinear* if it is contained in the same (horizontal, vertical or transversal) leaf of $M$, (respectively).
A *non-closed hexagonal configuration of $M$ with center $p_0 \in M$ and initial vertex $p_1 \in M$* is defined by an ordered non-collinear set of points $p_0, p_1, p_2, p_3, p_4, p_5, p_6$ of $M$, such that the triples of points $p_1, p_0, p_4$; $p_2, p_0, p_5$; $p_3, p_0, p_6$ and the pairs of points $p_i, p_{i+1}$, $(i = 1, 2, 3, 4, 5)$ are collinear.
A non-closed hexagonal configuration $(p_0, p_1, p_2, p_3, p_4, p_5, p_6)$ on $M$ *can be closed*, if the points $p_1$ and $p_6$ are (transversally) collinear.
A 3-web manifold $M$ is called *(locally) hexagonal* if (every point $q \in M$ has a neighbourhood $U_q$ such that) all non-closed hexagonal configurations of $M$ (contained in $U_q$) can be closed.

**Theorem 3.8 (W. Blaschke)** *A 2-dimensional 3-web manifold is locally hexagonal if and only if its Chern connection has vanishing curvature.*

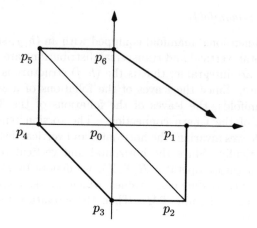

*Hexagonal configuration*

Figure 2.

*Proof.* Let be given a point $p_0$ of the manifold $M$. We consider a distinguished coordinate system $(x, y)$ centered at the point $p_0$ of the 3-web manifold $M$ in a neighbourhood $U$ of $p_0$. Then the coordinate functions $x$, $y$ and the function $f$ satisfy

$$f(0,0) = 0, \; f'_x(0,0) = f'_y(0,0) = 1,$$

$$f''_{xx}(0,0) = f''_{yy}(0,0) = f'''_{xxx}(0,0) = f'''_{yyy}(0,0) = 0.$$

Now, in this coordinate system we consider the hexagonal configuration centered at the point $p_0 = (0,0)$ and with initial vertex $p_1 = (x, 0)$. We put $p_2 = (x, u(x)), p_3 = (0, u(x)), p_4 = (u(x), 0), p_5 = (u(x), v(x)), p_6 = (0, v(x))$ and compute the Taylor polynomial of the function $v(x)$ up to order 3. Clearly we have $u(0) = v(0) = 0$. The functions $u(x)$, $v(x)$ satisfy the identities

$$f(x, u(x)) \equiv f(u(x), v(x)) \equiv 0.$$

Using the Implicit Function Theorem a direct computation gives:

$$\frac{du}{dx}(0) = -1, \quad \frac{d^2u}{dx^2}(0) = 2f''_{xy}(0,0),$$

$$\frac{d^3u}{dx^3}(0) = 3f'''_{xxy}(0,0) - 3f'''_{xyy}(0,0) - 6f''_{xy}(0,0)^2,$$

and

$$\frac{dv}{dx}(0) = 1, \quad \frac{d^2v}{dx^2}(0) = 0, \quad \frac{d^3v}{dx^3}(0) = 6\{f'''_{xyy}(0,0) - f'''_{xxy}(0,0)\}.$$

Hence we have

$$v(x) = x + \{f'''_{xyy}(0,0) - f'''_{xxy}(0,0)\}x^3 + o(3).$$

This means that the equation

$$\frac{1}{3!}\frac{d^3v}{dx^3}(0) = f'''_{xyy}(0,0) - f'''_{xxy}(0,0) = 0$$

is a necessary (infinitesimal) condition of the closeness of the hexagonal configuration at the point $p_0 = (0,0)$. Consequently, on a locally hexagonal 3-web manifold one has $f'''_{xyy}(0,0) - f'''_{xxy}(0,0) = 0$.
Hence we call the value $f'''_{xyy}(0,0) - f'''_{xxy}(0,0)$ the *measure of the non-closeness of the hexagonal configuration* and prove that it is equal to the unique scalar invariant of the curvature tensor for a 2-dimensional web manifold. Indeed, we put

$$X = \frac{\partial}{\partial x}, \quad Y = \frac{\partial}{\partial y}.$$

Taking into account that the Chern connection leaves invariant the horizontal and vertical distributions and using $[X,Y] = 0$ we can write

$$\nabla_X X = a(x,y)X, \quad \nabla_X Y = b(x,y)Y, \quad \nabla_Y X = c(x,y)X, \quad \nabla_Y Y = d(x,y)Y,$$

where $a(x,y), b(x,y), c(x,y), d(x,y)$ are smooth functions in the coordinate neighbourhood. Since $0 = T(X,Y) = \nabla_X Y - \nabla_Y X = bY - cX$, we obtain $b = c = 0$. Consequently the curvature tensorfield can be expressed as follows:

$$R(X,Y)X = r(X,Y)X = \{\nabla_X\nabla_Y - \nabla_Y\nabla_X\}X = -(\nabla_Y\nabla_X)X = -\frac{\partial a}{\partial y}X.$$

Hence we obtain

$$r(X,Y) = -\frac{\partial a}{\partial y}.$$

The function $a(x,y)$ can be computed from the defining relation of the Chern connection: $aX = hj[X, jX]$. The transversal curves are determined by the equations $f(x(t), y(t)) = const$ and hence a transversal direction $\dot{x}X + \dot{y}Y$ satisfies $f'_x\dot{x} + f'_y\dot{y} = 0$. Since the tensorfield $j$ is determined by the condition: for any vectorfield $Z$ the vectorfield $Z - jZ$ is transversal, we obtain

$$jX = f'_x(f'_y)^{-1}Y, \quad jY = f'_y(f'_x)^{-1}X.$$

Hence

$$a(x,y)X = hj\frac{\partial f_x'(f_y')^{-1}}{\partial x}Y = f_y'(f_x')^{-1}\frac{\partial f_x'(f_y')^{-1}}{\partial x}X$$

and

$$a(x,y) = (f_x')^{-1}f_{xx}'' - (f_y')^{-1}f_{xy}''.$$

This gives at the point $p_0 = (0,0)$ for the curvature

$$r_{(0,0)}(X,Y) = -\frac{\partial a(x,y)}{\partial y}(0,0) = f_{xyy}'''(0,0) - f_{xxy}'''(0,0).$$

But this value coincides with the measure of the non-closeness of the hexagonal configuration

$$r_{(0,0)}(U,V) = \frac{1}{6}\frac{d^3v}{dx^3}(0).$$

It follows that on a locally hexagonal 3-web on a 2-manifold $M$ the curvature tensorfield (and also the torsion tensorfield) of the Chern connection vanishes identically.

Hence by Theorem 2.14 any 2-dimensional locally hexagonal 3-web is parallelizable and therefore it is locally isomorphic to an affine web. □

**Corollary 3.9** *A 3-web on a 2-manifold is locally hexagonal if and only if it is parallelizable.*

*Any connected and simply connected locally hexagonal 3-web with complete Chern connection on a 2-manifold is web-isomorphic to the affine 3-net on the 2-plane.*

*The locally hexagonal 3-webs on 2-manifolds with complete Chern connection are web-isomorphic to the quotients $N/\Gamma$ of an affine 3-net $N$ on the 2-plane, where $\Gamma$ is a translation group acting freely and in a totally disconnected way on the affine plane $N$.*

## 3.4   Reidemeister configuration

In the previous section we showed that the curvature can be used as the measure of the non-closeness of hexagonal configurations. For dimension $n > 2$ the situation is more complicated, since the curvature tensorfield has more components. Now we want to give a deeper analysis of the geometric interpretation of the curvature tensorfield as measure of the non-closeness of classical configurations.

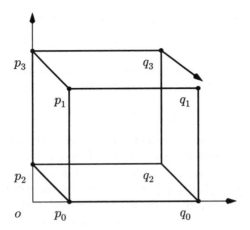

*Reidemeister configuration*

Figure 3.

**Definition 3.10** Let $\{p_0, q_0\}$, $\{p_1, q_1\}$, $\{p_2, q_2\}$, $\{p_3, q_3\}$ be horizontally collinear pairs of points such that the pairs $\{p_0, p_1\}, \{p_2, p_3\}, \{q_0, q_1\}, \{q_2, q_3\}$ are vertically collinear and the pairs $\{p_0, p_2\}, \{p_1, p_3\}, \{q_0, q_2\}$ are transversally collinear. Then the set of the points $\{p_0, q_0, p_1, q_1, p_2, q_2, p_3, q_3\}$ is called a *non-closed Reidemeister configuration*.

We say that a non-closed Reidemeister configuration $\{p_0, q_0, p_1, q_1, p_2, q_2, p_3, q_3\}$ *can be closed*, if the points $q_1$ and $q_3$ are transversally collinear.

We say that a 3-web manifold $M$ satisfies *(locally) the Reidemeister condition* if (every point $q \in M$ has a neighbourhood $U_q$ such that) all non-closed Reidemeister configurations of $M$ (contained in $U_q$) can be closed.

Let be given a point $o$ of the 3-web manifold $M$. Let $(\mathbf{x}, \mathbf{y}) = (x^1, ..., x^n; y^1, ..., y^n)$ be a distinguished coordinate system on a neighbourhood $U$ centered at the point $o \in U$, such that the point $o$ is given by the coordinates $(\mathbf{0}, \mathbf{0})$ and the connected components of the horizontal, vertical and transversal leaves have the equations $\mathbf{y} = \mathbf{0}$, $\mathbf{x} = \mathbf{0}$ and $\mathbf{f}(\mathbf{x}, \mathbf{y}) = const$, respectively, for a smooth mapping $\mathbf{f}$. In the distinguished coordinate system we have

$$\mathbf{f}(\mathbf{0}, \mathbf{0}) = \mathbf{0}, \ \mathbf{f}'_{\mathbf{x}}(\mathbf{0}, \mathbf{0}) = \mathbf{f}'_{\mathbf{y}}(\mathbf{0}, \mathbf{0}) = id,$$

$$\mathbf{f}''_{\mathbf{xx}}(\mathbf{0}, \mathbf{0}) = \mathbf{f}''_{\mathbf{yy}}(\mathbf{0}, \mathbf{0}) = \mathbf{f}'''_{\mathbf{xxx}}(\mathbf{0}, \mathbf{0}) = \mathbf{f}'''_{\mathbf{yyy}}(\mathbf{0}, \mathbf{0}) = \mathbf{0}.$$

Consequently the function $\mathbf{f}(\mathbf{x}, \mathbf{y})$ can be written in the form

$$\mathbf{f}(\mathbf{x}, \mathbf{y}) = \mathbf{x} + \mathbf{y} + \mathbf{q}(\mathbf{x}, \mathbf{y}) + \mathbf{r}(\mathbf{x}, \mathbf{x}, \mathbf{y}) + \mathbf{s}(\mathbf{x}, \mathbf{y}, \mathbf{y}) + \omega(\mathbf{x}, \mathbf{y}) \tag{2}$$

where $\mathbf{q}(\mathbf{x}, \mathbf{y})$ is bilinear, $\mathbf{r}(\mathbf{x}, \mathbf{y}, \mathbf{z})$ and $\mathbf{s}(\mathbf{x}, \mathbf{y}, \mathbf{z})$ are trilinear vector-valued functions of the distinguished coordinates and

$$\lim_{t \to 0} t^{-3} \cdot \omega(t\mathbf{x}, t\mathbf{y}) = 0.$$

The function $\mathbf{f}(\mathbf{x}, \mathbf{y})$ is the coordinate expression of the multiplication of the local loop associated with the 3-web in the distinguished coordinate system . Now we consider a Reidemeister configuration $\{p_0, q_0\}, \{p_1, q_1\}, \{p_2, q_2\}, \{p_3, q_3\}$ such that the intersection of the horizontal leaf through $p_0, q_0$ and of the vertical leaf through $p_2, p_3$ is the center $o$ of the distinguished coordinate system $U$ and all points $\{p_0, q_0\}, \{p_1, q_1\}, \{p_2, q_2\}, \{p_3, q_3\}$ are contained in $U$. We denote the coordinates of the points of this configuration in the following way: $o = (\mathbf{0}, \mathbf{0})$, $p_0 = (\mathbf{y}, \mathbf{0})$, $p_1 = (\mathbf{y}, \mathbf{z})$, $p_2 = (\mathbf{0}, \mathbf{y})$, $p_3 = (\mathbf{0}, \mathbf{f}(\mathbf{y}, \mathbf{z}))$, $q_0 = (\mathbf{f}(\mathbf{x}, \mathbf{y}), \mathbf{0})$, $q_1 = (\mathbf{f}(\mathbf{x}, \mathbf{y}), \mathbf{z})$, $q_2 = (\mathbf{x}, \mathbf{y})$, $q_3 = (\mathbf{x}, \mathbf{f}(\mathbf{y}, \mathbf{z}))$. Then

$$\Delta(\mathbf{x}, \mathbf{y}, \mathbf{z}) = \mathbf{f}(\mathbf{x}, \mathbf{f}(\mathbf{y}, \mathbf{z})) - \mathbf{f}(\mathbf{f}(\mathbf{x}, \mathbf{y}), \mathbf{z}) = 0$$

if and only if the Reidemeister configuration can be closed. Since the function $\mathbf{f}(\mathbf{x}, \mathbf{y})$ is the coordinate expression of the associated local loop multiplication, we obtain that the Reidemeister configuration can be closed (locally) if and only if the associated local loop multiplication is associative, i.e. if this local loop is a local group.

An easy calculation shows that this expression can be written in the form:

$$\Delta(\mathbf{x}, \mathbf{y}, \mathbf{z}) = \mathbf{q}(\mathbf{x}, \mathbf{q}(\mathbf{y}, \mathbf{z})) - \mathbf{q}(\mathbf{q}(\mathbf{x}, \mathbf{y}), \mathbf{z}) - \mathbf{r}(\mathbf{x}, \mathbf{y}, \mathbf{z}) -$$

$$-\mathbf{r}(\mathbf{y}, \mathbf{x}, \mathbf{z}) + \mathbf{s}(\mathbf{x}, \mathbf{y}, \mathbf{z}) + \mathbf{s}(\mathbf{x}, \mathbf{z}, \mathbf{y}) + \Omega(\mathbf{x}, \mathbf{y}, \mathbf{z}) \tag{3}$$

where

$$\lim_{t \to 0} t^{-3} \cdot \Omega(t\mathbf{x}, t\mathbf{y}, t\mathbf{z}) = 0.$$

**Theorem 3.11** *Let $u(t), v(t), w(t)$ be differentiable curves emanating from the point $o = u(0) = v(0) = w(0)$ and contained in the horizontal leaf through $o$. Then we have*

$$\lim_{t \to 0} t^{-2} \cdot (u(t) \circ v(t))/(v(t) \circ u(t)) = \lim_{t \to 0} t^{-2} \cdot (v(t) \circ u(t)) \backslash (u(t) \circ v(t)) \tag{4}$$

$$= T(\dot{u}(0), \dot{v}(0)) \tag{5}$$

*and*

$$\lim_{t \to 0} t^{-3} \cdot (u(t) \circ (v(t) \circ w(t)))/((u(t) \circ (v(t)) \circ w(t)) = \qquad (6)$$

$$= \lim_{t \to 0} t^{-3} \cdot ((u(t) \circ v(t)) \circ w(t)) \backslash (u(t) \circ (v(t) \circ w(t))) = \qquad (7)$$

$$= R(\dot{u}(0), j\dot{w}(0))\dot{v}(0),$$

*where $T$ and $R$ are the torsion and the curvature tensorfields of the Chern connection.*

*Proof.* First we investigate the expression (3). If $\mathbf{g}(\mathbf{x}, \mathbf{y})$ denotes the coordinate function of the right division $(u, v) \mapsto u/v$ in the distinguished coordinate system in the neighbourhood $U$ of the origin $o$ then we have

$$\mathbf{g}(\mathbf{x}, \mathbf{y}) = \mathbf{x} - \mathbf{y} + \mu(\mathbf{x}, \mathbf{y}),$$

where

$$\lim_{t \to 0} t^{-1} \cdot \mu(t\mathbf{x}, t\mathbf{y}) = 0.$$

Hence we have

$$\mathbf{g}(\mathbf{f}(\mathbf{x}, \mathbf{y}), \mathbf{f}(\mathbf{y}, \mathbf{z})) = \mathbf{q}(\mathbf{x}, \mathbf{y}) - \mathbf{q}(\mathbf{y}, \mathbf{x}) + \nu(\mathbf{x}, \mathbf{y}),$$

where

$$\lim_{t \to 0} t^{-2} \cdot \mu(t\mathbf{x}, t\mathbf{y}) = 0.$$

Similarly we can write

$$\mathbf{g}(\mathbf{f}(\mathbf{x}, \mathbf{f}(\mathbf{y}, \mathbf{z})), \mathbf{f}(\mathbf{f}(\mathbf{x}, \mathbf{y}), \mathbf{z}) =$$

$$= \Delta(\mathbf{x}, \mathbf{y}, \mathbf{z}) + \Lambda(\mathbf{x}, \mathbf{y}, \mathbf{z}) =$$

$$= \mathbf{q}(\mathbf{x}, \mathbf{q}(\mathbf{y}, \mathbf{z})) - \mathbf{q}(\mathbf{q}(\mathbf{x}, \mathbf{y}), \mathbf{z}) - \mathbf{r}(\mathbf{x}, \mathbf{y}, \mathbf{z}) - \mathbf{r}(\mathbf{y}, \mathbf{x}, \mathbf{z}) +$$

$$+ \mathbf{s}(\mathbf{x}, \mathbf{y}, \mathbf{z}) + \mathbf{s}(\mathbf{x}, \mathbf{z}, \mathbf{y}) + \Omega(\mathbf{x}, \mathbf{y}, \mathbf{z}) + \Lambda(\mathbf{x}, \mathbf{y}, \mathbf{z}),$$

where

$$\lim_{t \to 0} t^{-3} \cdot \Omega(t\mathbf{x}, t\mathbf{y}, t\mathbf{z}) = \lim_{t \to 0} t^{-3} \cdot \Lambda(\mathbf{x}, \mathbf{y}, \mathbf{z}) = 0.$$

On the other hand we can prove the following

**Proposition 3.12** *In a distinguished coordinate neighbourhood $U$ one has the expressions*

$$\mathbf{f}(\mathbf{x}, \mathbf{y}) - \mathbf{f}(\mathbf{y}, \mathbf{x}) = -T(\mathbf{y}, \mathbf{x}) + \omega(\mathbf{x}, \mathbf{y}),$$

*where*

$$\lim_{t \to 0} t^{-2} \cdot \omega(t\mathbf{x}, t\mathbf{y}) = 0,$$

*and*

$$\Delta(\mathbf{x}, \mathbf{y}, \mathbf{z}) = (\mathbf{f}(\mathbf{x}, \mathbf{f}(\mathbf{y}, \mathbf{z})) - \mathbf{f}(\mathbf{f}(\mathbf{x}, \mathbf{y}), \mathbf{z}) = R(\mathbf{x}, j\mathbf{z})\mathbf{y} + \Omega(\mathbf{x}, \mathbf{y}, \mathbf{z}),$$

*where*

$$\lim_{t \to 0} t^{-3} \cdot \Omega(t\mathbf{x}, t\mathbf{y}, t\mathbf{z}) = 0.$$

*Proof.* Let $X(\mathbf{x}, \mathbf{y}), Y(\mathbf{x}, \mathbf{y}), Z(\mathbf{x}, \mathbf{y})$ be constant vector-valued functions of the coordinate vectors $\mathbf{x}$ and $\mathbf{y}$ on the distinguished coordinate neighbourhood $U$, defining horizontal local tangent vectorfields of $M$. We want to express the values $T(X, Y) = \nabla_X Y - \nabla_Y X - [X, Y]$ of the torsion and $R(X, jY)Z = (\nabla_X \nabla_{jY} - \nabla_{jY} \nabla_X - \nabla_{[X, jY]})Z$ of the curvature at the point $o = (\mathbf{0}, \mathbf{0})$. For the computation we use the following fact: *If a horizontal vectorfield $X$ is given by a constant vector-valued function of the coordinate vectors $\mathbf{x}$ and $\mathbf{y}$ on the distinguished coordinate neighbourhood $U$ and $W$ is a vertical vectorfield then the vectorfield $[X, W]$ is vertical too.* It follows from the expression (1) of the covariant derivative that

$$T(X, Y) = j[X, jY] - j[Y, jX]$$

and

$$R(X, jY)Z =$$

$$= \nabla_X(h[jY, Z]) - \nabla_{jY}(hj[X, jZ]) - h[[X, jY], Z] = -h[jY, j[X, jZ]].$$

If $W$ is a vertical vectorfield which is given by a constant vector-valued function, then we obtain similarly

$$R(X, W)Z = -h[W, j[X, jZ]] = -[W, j[X, jZ]]. \tag{8}$$

We denote

$$j\left(\frac{\partial}{\partial x^i}\right) = \sum_i J_i^k \frac{\partial}{\partial y^k}, \quad j\left(\frac{\partial}{\partial y^i}\right) = \sum_i \tilde{J}_i^k \frac{\partial}{\partial x^k},$$

where the matrix $\tilde{J}_i^k$ is the inverse of the matrix $J_i^k$ since $j^2 = id$. Then we have

$$T((\frac{\partial}{\partial x^l}, \frac{\partial}{\partial x^m}) = T(\frac{\partial}{\partial x^l}, \sum_k J_m^k j(\frac{\partial}{\partial y^k})) =$$

$$= j \sum_h (\frac{\partial}{\partial x^l}(J_m^h) - \frac{\partial}{\partial x^m}(J_l^h)) \frac{\partial}{\partial y^h} = \sum_h (\frac{\partial}{\partial x^l}(J_m^h) - \frac{\partial}{\partial x^m}(J_l^h)) \sum_r \tilde{J}_h^r \frac{\partial}{\partial x^r}.$$

(9)

Moreover

$$R(\frac{\partial}{\partial x^l}, j(\frac{\partial}{\partial x^i})) \frac{\partial}{\partial x^m} = R(\frac{\partial}{\partial x^l}, \sum_k J_i^k \frac{\partial}{\partial y^k})) \frac{\partial}{\partial x^m} = \sum_k J_i^k R(\frac{\partial}{\partial x^l}, \frac{\partial}{\partial y^k}) \frac{\partial}{\partial x^m} =$$

$$= -\sum_k J_i^k [\frac{\partial}{\partial y^k}, j[\frac{\partial}{\partial x^l}, \sum_h J_m^h \frac{\partial}{\partial y^h}]] = -\sum_k J_i^k [\frac{\partial}{\partial y^k}, j(\sum_h \frac{\partial}{\partial x^l}(J_m^h) \frac{\partial}{\partial y^h})] =$$

$$-\sum_{k,h,q} J_i^k [\frac{\partial}{\partial y^k}, \frac{\partial}{\partial x^l}(J_m^h) \tilde{J}_h^q \frac{\partial}{\partial x^q}] = -\sum_{k,h,q} J_i^k \frac{\partial}{\partial y^k}\{\frac{\partial}{\partial x^l}(J_m^h) \tilde{J}_h^q\} \frac{\partial}{\partial x^q} =$$

$$= -\sum_{k,h,q} J_i^k \{\frac{\partial^2}{\partial y^k \partial x^l}(J_m^h) \tilde{J}_h^q + \frac{\partial}{\partial x^l}(J_m^h) \frac{\partial}{\partial y^k}(\tilde{J}_h^q)\} \frac{\partial}{\partial x^q}.$$

(10)

Now we compute the tensorfield $j$. The horizontal component of the tangent vector $(\dot{\mathbf{x}}, \dot{\mathbf{y}})$ of a curve $(\mathbf{x}(t), \mathbf{y}(t))$ is $(\dot{\mathbf{x}}, \mathbf{0})$ and hence the transversal vector with this horizontal component is $(\dot{\mathbf{x}}, -j\dot{\mathbf{x}})$. If this curve lies on a transversal leaf it satisfies $\mathbf{f}(\mathbf{x}(t), \mathbf{y}(t)) = const$ and hence its tangent vector satisfies the equations $\mathbf{f'_x} \cdot \dot{\mathbf{x}} + \mathbf{f'_y} \cdot \dot{\mathbf{y}} = 0$ and $\dot{\mathbf{y}} = -j\dot{\mathbf{x}}$. Thus the tensorfield $j$ is determined by the relation

$$(\mathbf{f'_x} - \mathbf{f'_y} \cdot j) \cdot \dot{\mathbf{x}} = 0$$

for any horizontal vector $\dot{\mathbf{x}}$. It follows that the coordinate functions $J_i^k$ and $\tilde{J}_i^k$ of the tensorfield $j$ satisfy

$$\frac{\partial f^i}{\partial x^k} = \frac{\partial f^i}{\partial y^m} J_k^m, \qquad \frac{\partial f^i}{\partial x^k} \tilde{J}_m^k = \frac{\partial f^i}{\partial y^m},$$

(11)

where $f^1(x^1, ..., x^n; y^1, ..., y^n)) = f^1(\mathbf{x}, \mathbf{y}), ..., f^n(x^1, ..., x^n; y^1, ..., y^n) = f^n(\mathbf{x}, \mathbf{y})$ are the coordinate functions of the smooth mapping $\mathbf{f}$. Putting $(\mathbf{x}, \mathbf{y}) = (\mathbf{0}, \mathbf{0})$ and using $\mathbf{f'_x} = id$, $\mathbf{f'_y} = id$ we obtain that $J_k^m = \delta_k^m$ and

$\tilde{J}^k_m = \delta^k_m$, where $\delta^m_k$ is the identity matrix. The derivation of the equations (10) gives at $(\mathbf{x}, \mathbf{y}) = (\mathbf{0}, \mathbf{0})$:

$$\frac{\partial^2 f^i}{\partial x^l \partial y^k} + \frac{\partial}{\partial x^l}(J^i_k) = 0, \quad \frac{\partial^2 f^i}{\partial x^k \partial y^p} - \frac{\partial}{\partial y^p}(J^i_k) = 0, \quad \frac{\partial^2 f^i}{\partial x^m \partial y^k} + \frac{\partial}{\partial y^k}(\tilde{J}^i_m) = 0.$$

(12)

Hence we obtain from the equation (7) at the point $(\mathbf{x}, \mathbf{y}) = (\mathbf{0}, \mathbf{0})$:

$$T(\frac{\partial}{\partial x^l}, \frac{\partial}{\partial x^m}) = \sum_i (-\frac{\partial^2 f^i}{\partial x^l \partial y^m} + \frac{\partial^2 f^i}{\partial x^m \partial y^l}) \frac{\partial}{\partial x^i}.$$

Using the Taylor formula (1) of the function $\mathbf{f}(\mathbf{x}, \mathbf{y})$ at the point $(\mathbf{0}, \mathbf{0})$ we can write this equation in the form

$$T(\mathbf{x}, \mathbf{y}) = \mathbf{q}(\mathbf{y}, \mathbf{x}) - \mathbf{q}(\mathbf{x}, \mathbf{y})$$

and hence we obtain

$$\mathbf{f}(\mathbf{x}, \mathbf{y}) - \mathbf{f}(\mathbf{y}, \mathbf{x}) = -T(\mathbf{y}, \mathbf{x}) + \omega(\mathbf{x}, \mathbf{y}),$$

where

$$\lim_{t \to 0} t^{-2} \cdot \omega(t\mathbf{x}, t\mathbf{y}) = 0.$$

The other relation for the torsion can be proved similarly. Now we compute the second derivation of (10) at $(\mathbf{x}, \mathbf{y}) = (\mathbf{0}, \mathbf{0})$:

$$\frac{\partial^3 f^i}{\partial x^l \partial x^k \partial y^p} = \frac{\partial^3 f^i}{\partial x^l \partial y^k \partial y^p} + \frac{\partial^2 f^i}{\partial x^l \partial y^m} \frac{\partial}{\partial y^p}(J^m_k) + \frac{\partial^2}{\partial x^l \partial y^p}(J^i_k) = 0. \quad (13)$$

Using the equations (9),(11) and (12) we obtain at $(\mathbf{x}, \mathbf{y}) = (\mathbf{0}, \mathbf{0})$:

$$R(\frac{\partial}{\partial x^l}, j(\frac{\partial}{\partial x^i})) \frac{\partial}{\partial x^m} = -\sum_{k,h,q} \{\frac{\partial^2}{\partial y^i \partial x^l}(J^q_m) + \frac{\partial}{\partial x^l}(J^h_m) \frac{\partial}{\partial y^k}(\tilde{J}^q_h)\} \frac{\partial}{\partial x^q} =$$

$$-\sum_{k,h,q} (\frac{\partial^3 f^q}{\partial x^l \partial x^m \partial y^i} - \frac{\partial^3 f^q}{\partial x^l \partial y^m \partial y^i} - \frac{\partial^2 f^q}{\partial x^l \partial y^r} \frac{\partial^2 f^r}{\partial x^m \partial y^i} + \frac{\partial^2 f^h}{\partial x^l \partial y^m} \frac{\partial^2 f^q}{\partial x^h \partial y^i}).$$

We can write this equation at the point $(\mathbf{0}, \mathbf{0})$ in the form

$$R(\mathbf{x}, j\mathbf{z})\mathbf{y} = \mathbf{q}(\mathbf{x}, \mathbf{q}(\mathbf{y}, \mathbf{z})) - \mathbf{q}(\mathbf{q}(\mathbf{x}, \mathbf{y}), \mathbf{z}) - \mathbf{r}(\mathbf{x}, \mathbf{y}, \mathbf{z}) -$$

$$-\mathbf{r}(\mathbf{y}, \mathbf{x}, \mathbf{z}) + \mathbf{s}(\mathbf{x}, \mathbf{y}, \mathbf{z}) + \mathbf{s}(\mathbf{x}, \mathbf{z}, \mathbf{y}).$$

Consequently we obtain that the expression

$$\Delta(\mathbf{x}, \mathbf{y}, \mathbf{z}) = R(\mathbf{x}, j\mathbf{y})\mathbf{z} + \Omega(\mathbf{x}, \mathbf{y}, \mathbf{z})$$

is a measure of the non-closeness of the Reidemeister configuration, where

$$\lim_{t \to 0} t^{-3} \cdot \Omega(t\mathbf{x}, t\mathbf{y}, t\mathbf{z}) = 0.$$

□

The result for the expression (6) can be proved similarly. Now, from the previous consideration and from this proposition the assertion of the theorem follows. □

**Definition 3.13** The bilinear operation $[\mathbf{x}, \mathbf{y}] = -T(\mathbf{x}, \mathbf{y})$ and the trilinear operation $\langle \mathbf{x}, \mathbf{y}, \mathbf{z} \rangle = R(\mathbf{x}, j\mathbf{z})\mathbf{y}$ on the horizontal tangent space $T_o^h L$ at $o \in M$ of a 3-web are called the *commutator* and the *associator*, respectively, of the local loop $L$ centered at the point $o \in M$ and associated with the 3-web $M$.

## 3.5 Bol, Moufang and hexagonal webs

**Definition 3.14** A non-closed Reidemeister configuration

$$\{p_0, \ q_0, \ p_1, \ q_1, \ p_2, \ q_2, \ p_3, \ q_3\}$$

is called a *non-closed vertical Bol configuration*, if the points $\{p_0, \ p_1, \ q_2, \ q_3\}$ are vertically collinear. The non-closed Reidemeister configuration

$$\{p_0, \ q_0, \ p_1, \ q_1, \ p_2, \ q_2, \ p_3, \ q_3\}$$

is called a *non-closed horizontal Bol configuration*, if the points $\{p_1, \ q_1, \ p_2, \ q_2\}$ are horizontally collinear. We say that a non-closed Bol configuration *can be closed*, if it can be closed as a Reidemeister configuration.

A 3-web manifold $M$ satisfies *(locally) the vertical (horizontal) Bol condition*, if (every point $q \in M$ has a neighbourhood $U_q$ such that) all non-closed vertical (horizontal) Bol configurations of $M$ (contained in $U_q$) can be closed.

A 3-web manifold $M$ satisfies *(locally) the Moufang condition* if it (locally) satisfies both the vertical and the horizontal Bol conditions.

We notice that if the points $p_1$ and $q_2$ coincide in a non-closed Reidemeister configuration then we obtain a hexagonal configuration.

Since the horizontal and vertical Bol conditions, the Moufang condition and the hexagonal condition can be obtained from the Reidemeister condition by specialization of the Reidemeister configuration, we get the following assertions as consequences of the description of 3-webs satisfying the Reidemeister condition:

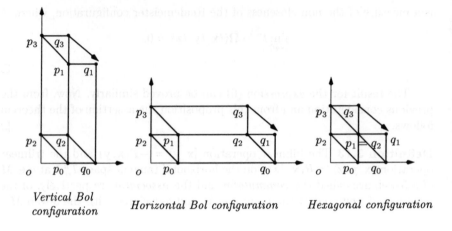

Figure 4.

**Theorem 3.15** *A (local) closure condition for a 3-web M is equivalent with a weak associative identity for any associated local loops L:*

| Configuration: | Identity: |
|---|---|
| Reidemeister | $(u \circ v) \circ w = u \circ (v \circ w)$ |
| Moufang | $\begin{cases} (u \circ u) \circ w = u \circ (u \circ w) \\ (u \circ w) \circ w = u \circ (w \circ w), \end{cases}$ |
| vertical Bol | $(u \circ u) \circ w = u \circ (u \circ w),$ |
| horizontal Bol | $(u \circ w) \circ w = u \circ (w \circ w),$ |
| hexagonal | $(w \circ w) \circ w = w \circ (w \circ w)$ |

*for all $u, v, w \in L$.*

Using Theorem 3.11 we obtain the following

**Theorem 3.16** *The local closure conditions for a 3-web manifold M imply the following identities for the curvature tensorfield of its Chern connection:*

| Configuration: | Curvature identity: |
|---|---|
| Reidemeister | $R(X, jY)Z = 0$ |
| Moufang | $R(X, jX)Z = R(X, jZ)Z = R(X, jY)X = 0,$ |
| vertical Bol | $R(X, jZ)X = 0,$ |
| horizontal Bol | $R(X, jZ)Z = 0,$ |
| hexagonal | $R(X, jX)X = 0$ |

*for any horizontal vectorfields $X, Y, Z$ on $M$.*

*Remark.* From Theorem 2.10 we know that the curvature identity $R(X, jY)Z = 0$ is necessary and sufficient for the local Reidemeister closure condition.

## 4  Special classes of 3-webs

This section is devoted to the proof of some fundamental results of the Akivis' theory of 3-webs, asserting that the curvature identities, which are the infinitesimal versions of the hexagonal, Bol or Moufang conditions for 3-webs by Theorem 3.16, give infinitesimal *characterizations* of the classical closure conditions. First we investigate the class of transversally geodesic 3-webs, for which there exists a maximal family of 2-dimensional subwebs. It follows from their curvature characterization that the 3-webs satisfying the necessary curvature condition of locally hexagonal 3-webs have a maximal family of 2-dimensional subwebs. Hence the investigation of such 3-webs can be reduced to the 2-dimensional case. Using Blaschke's theorem we can obtain that these 3-webs are locally hexagonal in the higher dimensional case too.

For the investigation of 3-webs satisfying the curvature condition of Bol 3-webs we use a coordinate expression of the curvature tensor in a distinguished coordinate system, from which it follows that the connection coefficients of the torsionless connection having the same geodesics as the induced connection of the horizontal leaves do not depend on the vertical coordinates and hence the vertical projection between horizontal leaves is an affine map. Using the configuration properties of the existing maximal family of 2-dimensional subwebs in this case too, one can prove the equivalence of the curvature condition with the Bol closure condition.

Finally we describe the geometry of 3-webs having torsionless Chern connection.

### 4.1  Subwebs and transversally geodesic 3-webs

**Definition 4.1** Suppose that a 3-web structure is given on the submanifold $S$ of the 3-web manifold $M$. $S$ is called a *subweb* of the 3-web $M$, or we say that *the 3-web structure on the submanifold $S$ is induced by the 3-web structure on $M$* if the leaves of $S$ are intersections of $S$ with the leaves of the 3-web $M$.

$K$ is a *subnet* of a 3-net manifold $M$ if it is a subweb of $M$.

**Definition 4.2** A 2-dimensional subspace in a tangent space $T_p M$ of a web manifold $M$ is called *transversal* if it intersects the horizontal (or vertical or transversal) subspace of $T_p M$ and is invariant with respect to the operator $j_p : T_p M \to T_p M$. A 2-surface $\phi : S \to M$ immersed into the web manifold $M$ is called *transversally geodesic* if the images of its tangent planes $\phi_* T_q S \subset T_{\phi(q)} M$ are transversal subspaces.

**Proposition 4.3** *A 2-dimensional submanifold of a 3-web manifold $M$ has an induced subweb structure if and only if it is a transversally geodesic 2-surface.*

*Proof.* If the immersed 2-surface $\phi : S \to M$ is transversally geodesic then it is clear that the tensorfields $h, v, j$ associated with the 3-web on $M$ can be restricted to $\phi_* T_q S$, consequently the transversally geodesic 2-surface $S$ has an induced 3-web structure.

On the other hand, if $S$ is an immersed two-dimensional subweb and $p \in S$ is a point on $S$, then its horizontal, vertical and transversal tangent vectors are contained in the tangent plane $T_p S$ and this plane is invariant with respect to the operator $j_p$. It follows that $S$ is transversally geodesic. $\qquad\square$

In what follows for the sake of simplicity we shall consider the immersed transversally geodesic surfaces as subsets in the manifold $M$.

**Proposition 4.4** *The transversally geodesic 2-surfaces of a web manifold $M$ are autoparallel submanifolds with respect to the Chern connection $\nabla$.*

*Proof.* Let $U$ and $jU$ be horizontal and vertical vectorfields defined on a neighbourhood of a transversally geodesic 2-surface $S \subset M$. An arbitrary tangent vectorfield on $S$ can be written in the form $\alpha U + \beta jU$, where $\alpha$ and $\beta$ are functions on $S$. One can calculate

$$\nabla_{\alpha U + \beta jU}(\lambda U + \mu jU) = (\alpha U \lambda + \beta jU \lambda)U + (\alpha U \mu + \beta jU \mu)jU +$$

$$+ \alpha \lambda \nabla_U U + \alpha \mu \nabla_U(jU) + \beta \lambda \nabla_{jU} U + \beta \mu \nabla_{jU}(jU).$$

We have

$$\nabla_U U = hj[U, jU], \qquad \nabla_U(jU) = v[U, jU],$$

$$\nabla_{jU} U = h[jU, U], \qquad \nabla_{jU} = vj[jU, U].$$

Since $U$ and $jU$ span an integrable 2-plane field, we get from Frobenius' theorem $[U, jU] = \rho U + \sigma jU$, that is $\nabla_{\alpha U + \beta jU}(\lambda U + \mu jU)$ is contained in the tangent 2-plane $T_p S$. $\qquad\square$

**Corollary 4.5** *The leaves of the subweb induced on a transversally geodesic 2-surface S of a 3-web manifold M are geodesics with respect to the Chern connection.*

*Proof.* The assertion follows from the fact that the leaves of a 3-web are autoparallel submanifolds with respect to the Chern connection. ☐

**Definition 4.6** A 3-web on the manifold $M$ is called *transversally geodesic* if for any horizontal vectors $U_p \in T_p^{(h)} M, p \in M$ there exists a transversally geodesic 2-surface tangent to the 2-plane spanned by the vectors $U_p, jU_p$.

**Theorem 4.7** *Let be given a 3-web manifold $M$. The following conditions are equivalent:*
*(i) The 3-web structure on $M$ is transversally geodesic.*
*(ii) The 3-web manifold $M$ has a maximal family of 2-dimensional subwebs.*
*(iii) There exists a (2,0)-tensorfield $q(X,Y)$ on $M$ such that the curvature tensorfield $R$ of the Chern connection of $M$ satisfies*

$$\sigma\{q(X,Y)Z\} = \sigma\{R(X,jY)Z\}$$

*for all horizontal vectors $X, Y, Z \in T_p^{(h)} M, p \in M$, where $\sigma$ denotes the summation for all permutations of the variables $X, Y, Z$.*

*Proof.* The transversally geodesic submanifolds are totally geodesic and coincide with the 2-dimensional subwebs. Hence if two 2-dimensional subwebs have the same transversal tangent subspace, then they coincide. It follows that the conditions (i) and (ii) are equivalent. Let $S$ be a transversally geodesic 2-surface in $M$. We suppose that the horizontal vectorfield $U$ on $S$ is normalized by the condition $\nabla_U U = 0$. Then its integral curves are affine parameterized horizontal geodesics on $S$. If we consider a vertical geodesic on $S$ then with the help of its parameter we can describe a 1-parameter family of horizontal geodesics which is a geodesic variation on $M$. The corresponding infinitesimal variation vectorfield $W$ satisfies the Jacobi differential equation

$$\nabla_U \nabla_U W + \nabla_U(T(W,U)) + R(W,U)U = 0$$

along a horizontal geodesic (cf. Kobayashi-Nomizu [14,15], *Theorem 1.2* in Chapter VIII). We denote $HW = \alpha U, VW = \beta jU$ where $\alpha$ and $\beta \neq 0$ are real functions, then the Jacobi equation can be written in the form

$$\nabla_U \nabla_U(\alpha U + \beta jU) = \beta R(U, jU)U,$$

since $T(U,U) = T(jU,U) = 0$ and $R(\alpha U + \beta jU, U)U = \beta R(U, jU)U$. Using $\nabla_U U = \nabla_U jU = 0$ we obtain the equivalent equation $(U(U\alpha))U +$

$(U(U\beta))jU = \beta R(U,jU)U$. Since the right hand side of this equation is horizontal it can be written as

$$(U(U\alpha))U = \beta R(U,jU)U,$$

$$U(U\beta) = 0.$$

It is easy to see that if the functions $\alpha$ and $\beta$ satisfy these equations then $\alpha U + \beta jU$ is a Jacobi vectorfield along the horizontal geodesic integral curve of the vectorfield $U$.

If the 3-web is transversally geodesic then it follows from the preceding equations that for any horizontal vector $U_p \in T_p^{(h)}M, p \in M$ there exists a scalar $f(U_p)$ satisfying

$$f(U_p)U_p = R_p(U_p, jU_p)U_p.$$

Since the right hand side of this equation is a cubic form in $U_p$ the function $f$ is a quadratic form in $U_p$. Hence it can be written in the form $f(U) = q(U,U)$, where $g$ is a (2,0)-tensorfield on $M$. Thus for transversally geodesic 3-webs we obtain the identity

$$R(U,jU)U = q(U,U)U$$

for horizontal vectorfields $U$ on $M$, which is equivalent to the condition (iii) in the *Theorem*.

Now we suppose that (iii) is satisfied. Let be given $p \in M, U_p \in T_p^{(h)}M$ and let $\gamma(w), w \in I \subset \mathbb{R}$ be the vertical geodesic with the initial value $\gamma(0) = p, \gamma'(0) = j_p U_p$. For a given $w \in I$ we consider the horizontal geodesic $\tilde{\gamma}_w(u)$ satisfying $\tilde{\gamma}_w(0) = \gamma(w)$ and $\tilde{\gamma}'(0) = j_{\gamma(w)}\gamma'(0)$. The geodesics $\tilde{\gamma}_w(u)$ define in the neighbourhood $(u,w) \in I \times I \subset R^2$ a 2-surface $\phi(u,w)$ tangent to the transversal 2-plane spanned by the vectors $U_p$ and $jU_p$. The tangent vectorfields $U(u,w) = \phi_* \frac{\partial}{\partial u}$ and $W(u,w) = \phi_* \frac{\partial}{\partial w}$ satisfy

$$\nabla_U U = 0, \qquad [U,W] = 0,$$

and along the vertical geodesic $\gamma(w)$

$$\nabla_W W = 0, \qquad \nabla_W U = \nabla_W jW = j\nabla_W W = 0.$$

We fix a $w_o \in I$. The vectorfield $W$ is a Jacobi vectorfield along the horizontal geodesic $\tilde{\gamma}_{w_o}(u)$. Hence it satisfies

$$\nabla_U \nabla_U W + \nabla_U(T(W,U)) + R(W,U)U = 0$$

with the initial values $W_{\gamma_{w_o}} = (jU)_{\gamma_{w_o}}$ and $(\nabla_U W)_{\gamma_{w_o}} = (T(U,W) + \nabla_W U + [U,W])_{\gamma_{w_o}} = 0$, since $(T(U,W))_{\gamma_{w_o}} = T_{\gamma_{w_o}}(j\gamma'_{w_o}, \gamma'_{wo}) = 0$.

On the other hand, if $\alpha(u)$ is a real function satisfying the differential equation

$$\frac{d^2\alpha}{du^2}(u) = g(U_{\tilde{\gamma}_{w_o}}, U_{\tilde{\gamma}_{w_o}})$$

with initial values $\alpha(0) = 0$, $\frac{d\alpha}{du}(0) = 0$, then the vectorfield

$$W_{\tilde{\gamma}_{w_o}(u)} = \alpha(u)U_{\tilde{\gamma}_{w_o}} + (jU)_{\tilde{\gamma}_{w_o}}$$

is a Jacobi vectorfield along the horizontal geodesic $\tilde{\gamma}_{w_o}(u)$. Since the tangent plane of the 2-surface $\phi(u,w)$ at the point $(u,w)$ is spanned by the horizontal vector $U(u,w)$ and by the value of the Jacobi vectorfield $W$ along the horizontal geodesic through $\phi(u,w) \in M$ and the Jacobi vectorfield $W$ is a linear combination of the vectorfields $U$ and $jU$, the 2-surface $\phi(u,v)$ is transversally geodesic. It follows from this construction that the 3-web on $M$ is transversally geodesic. $\qquad \square$

## 4.2   Hexagonal 3-webs

**Theorem 4.8** *Let be given a 3-web manifold $M$. The following conditions are equivalent: (i) The 3-web on $M$ satisfies the local hexagonal condition.*
*(ii) For any point $p \in M$ and any horizontal tangent vector $U_p \in T_p^{(h)}M$ there exists a parallelizable subweb $S$ tangent to $U_p$, that is the 3-web manifold $M$ has a maximal family of parallelizable subwebs.*
*(iii) The curvature tensorfield $R$ of the Chern connection of $M$ satisfies*

$$\sigma\{R(X, jY)Z\} = 0$$

*for all horizontal vectors $X, Y, Z \in T_p^{(h)}, p \in M$, where $\sigma$ denotes the summation for all permutations of the variables $X, Y, Z$.*

*Proof.* Let $M$ be a locally hexagonal 3-web manifold. Then according to Theorem 3.13 the curvature tensorfield of the Chern connection satisfies $R(X, jX)X = 0$ identically. Using Theorem 4.7 from this it follows that the 3-web on $M$ is transversally geodesic. Since the transversally geodesic 2-surfaces are autoparallel submanifolds, the curvature of the Chern connection with respect to the induced 3-web on a transversally geodesic surface coincides with the restriction of the curvature tensorfield of the Chern connection on $M$. Consequently the curvature tensorfield of the induced 3-webs on transversally geodesic 2-surfaces vanishes identically. Hence the induced 3-web on the transversally geodesic 2-surfaces is parallelizable and we obtain that (i) implies the equivalent conditions (ii). Since the identity $R(X, jX)X = 0$ for all horizontal vectors $X$ is equivalent to the condition (iii) we obtain that (ii) and (iii) are equivalent.

Let $p_0$ be a point in a 3-web manifold $M$ having the property (ii), or equivalently for which the curvature tensorfield of the Chern connection satisfies the identity (iii) . Let $U$ be a neighbourhood of $p_0$ in $M$ which is simply covered by geodesics emanating from $p_0$. We show that any non-closed hexagonal configuration centered at $p_0$ and with initial vertex $p_1 \in U$ can be closed. According to Theorem 4.7 the 3-web on $M$ is transversally geodesic, hence the points $p_0$ and $p_1 \in U$ can be connected by a horizontal geodesic $\gamma$ and hence the non-closed hexagonal configuration centered at $p_0$ and with initial vertex $p_1 \in U$ is contained in the transversally geodesic 2-surface tangent to the 2-plane spanned by the tangent vector $\mathbf{u}$ of the horizontal geodesic $\gamma$ and its image $j\mathbf{u}$ by the map $j$. The condition (ii) implies that the induced 3-webs on transversally geodesic 2-surfaces are parallelizable and hence satisfy the local hexagonal condition too. It follows that on all horizontal geodesics emanating from $p_0$ there is an open segment containing $p_0$ such that if the initial vertex $p_1 \in U$ of a non-closed hexagonal configuration belongs to this geodesic segment then the configuration can be closed. Hence the 3-web on $M$ is locally hexagonal. $\qquad\qquad\square$

### 4.3  Bol and Moufang 3-webs

**Definition 4.9** A 3-web satisfying the vertical local Bol condition is called a *Bol 3-web*. If it satisfies both the horizontal and the vertical local Bol conditions then it is called a *Moufang 3-web*.

**Theorem 4.10** *The curvature tensorfield of the Chern connection of a 3-web satisfies $R(X, jZ)X = 0$ for any horizontal vectorfields $X$, $Y$ if and only if it is a Bol 3-web.*

*Proof.* According to Theorem 3.16 the curvature tensorfield of a Bol 3-web satisfies the identity $R(X, jZ)X = 0$. Conversely, if this identity is satisfied then the 3-web is locally hexagonal (cf. Theorem 4.8 (iii)) and hence transversally geodesic too.

Let be given a configuration $\{p_0, q_0, p_1, q_1, p_2, q_2, p_3, q_3\}$ contained in a distinguished coordinate neighbourhood and such that the pairs of points $\{p_0, q_0\}$, $\{p_1, q_1\}$, $\{p_0, q_0\}$, $\{p_2, q_2\}$, $\{p_3, q_3\}$ are horizontally collinear and the pair $\{q_0, q_1\}$ and the quadruple $\{p_0, q_2, p_1, q_3\}$ are vertically collinear. Moreover we assume that the pairs $\{p_0, q_0\}$ and $\{p_1, q_1\}$ contained in the horizontal leaves

$L$ and $L'$, respectively, can be connected by horizontal geodesics. Clearly, the local vertical Bol condition is equivalent to the assumption that every configuration of this type has the property that the pair $\{p_0, p_3\}$ is vertically collinear.

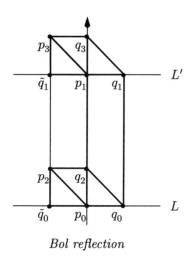

*Bol reflection*

Figure 5.

Now we consider the quadruples $p_0$, $q_0$, $p_2$, $q_2$ and $p_1$, $q_1$, $p_3$, $q_3$ in this configuration. Since the 3-web on $M$ is transversally geodesic, these quadruples are contained in transversally geodesic submanifolds. But the transversally geodesic submanifolds are parallelizable and the connection induced by the Chern connection on these transversally geodesic submanifolds is the canonical flat connection of the induced local affine 3-net structure. We denote by $\tilde{q}_0$ the intersection point of the horizontal geodesic containing the points $p_0$, $q_0$ with the vertical geodesic through $p_2$ and by $\tilde{q}_1$ the intersection point of the horizontal geodesic containing $p_1$, $q_1$ with the vertical geodesic through $p_3$. These intersection points exist if the pairs $p_0$, $q_0$ and $p_1$, $q_1$ are sufficiently close on the locally affine 3-nets on the transversally geodesic submanifolds. Since the locally affine 3-net consists of 3 families of parallel lines on the (local) affine plane structure given by the flat induced connection, for fixed centres $p_0$ and $p_1$ the mappings $q_0 \mapsto \tilde{q}_0$ and $q_1 \mapsto \tilde{q}_1$ coincide with the geodesic reflections of the horizontal leaves $L$ and $L'$, with respect the centres $p_0$ and $p_1$, respectively.

**Lemma 4.11** *Let $\nabla^L$ and $\nabla^{L'}$ denote the torsionless connections on the horizontal leaves $L$ and $L'$, respectively, which have the same geodesics as the connections induced by the Chern connection of a 3-web satisfying $R(X, jZ)X = 0$ for any horizontal vectorfields $X$, $Y$. The projection $\pi^{(v)} : L \to L'$ between the horizontal leaves $L$ and $L'$ along the vertical leaves is an affine map with respect to the connections $\nabla^L$ and $\nabla^{L'}$.*

*Proof.* Let $(\mathbf{x}, \mathbf{y}) = (x^1, ..., x^n; y^1, ..., y^n)$ be a distinguished coordinate system in a neighbourhood $U$ of the 3-web manifold $M$. We assume that the horizontal leaves $L \cap U$ and $L' \cap U$ in this neighbourhood $U$ are given by the equations $\mathbf{y} = \mathbf{y}_0$ and $\mathbf{y} = \mathbf{y}_1$, respectively. Then the coordinate function of the projection $\pi^{(v)} : L \to L'$ is the map $(\mathbf{x}, \mathbf{y}_0) \to (\mathbf{x}, \mathbf{y}_1)$ and the tangent map $\pi_*^{(v)}$ is given by the coordinate map

$$\frac{\partial}{\partial x^i}\Big|_{(\mathbf{x}, \mathbf{y}_0)} \mapsto \frac{\partial}{\partial x^i}\Big|_{(\mathbf{x}, \mathbf{y}_1)}.$$

If we denote

$$\nabla_{\frac{\partial}{\partial x^j}} \frac{\partial}{\partial x^k} = \sum_{i=1}^{n} \Gamma_j{}^i{}_k(\mathbf{x}, \mathbf{y}) \frac{\partial}{\partial x^i}$$

then we have

$$\nabla^L_{\frac{\partial}{\partial x^j}} \frac{\partial}{\partial x^k} = \sum_{i=1}^{n} (\Gamma_j{}^i{}_k(\mathbf{x}, \mathbf{y}_0) + \Gamma_k{}^i{}_j(\mathbf{x}, \mathbf{y}_0)) \frac{\partial}{\partial x^i}$$

and

$$\nabla^{L'}_{\frac{\partial}{\partial x^j}} \frac{\partial}{\partial x^k} = \sum_{i=1}^{n} (\Gamma_j{}^i{}_k(\mathbf{x}, \mathbf{y}_1) + \Gamma_k{}^i{}_j(\mathbf{x}, \mathbf{y}_1)) \frac{\partial}{\partial x^i}.$$

According to the equation (3.5) we have

$$R(\frac{\partial}{\partial x^j}, \frac{\partial}{\partial y^h}) \frac{\partial}{\partial x^k} = -[\frac{\partial}{\partial y^h}, \nabla_{\frac{\partial}{\partial x^j}} \frac{\partial}{\partial x^k}] = \sum_{i=1}^{n} \frac{\partial}{\partial y^h} (\Gamma_j{}^i{}_k(\mathbf{x}, \mathbf{y}_1)) \frac{\partial}{\partial x^i}.$$

Hence, the condition $R(X, jZ)X = 0$ for any horizontal vectorfields $X$, $Y$ implies

$$0 = R(\frac{\partial}{\partial x^j}, \frac{\partial}{\partial y^h}) \frac{\partial}{\partial x^k} + R(\frac{\partial}{\partial x^k}, \frac{\partial}{\partial y^h}) \frac{\partial}{\partial x^j} = \sum_{i=1}^{n} \frac{\partial}{\partial y^h} (\Gamma_j{}^i{}_k + \Gamma_k{}^i{}_j) \frac{\partial}{\partial x^i},$$

which means that the coefficients of the connections $\nabla^L$ and $\nabla^{L'}$ do not depend on the changing of the coordinate $\mathbf{y}$. Since the map $\pi^{(v)} : L \to L'$ is given by changing the coordinate $\mathbf{y}$, we obtain the lemma. $\square$

Now, for the proof of the theorem we notice that $\pi^{(v)} : L \to L' : p_0 \mapsto p_1$ and $\pi^{(v)} : q_0 \mapsto p_1$. Since the points $\tilde{q}_0$ and $\tilde{q}_1$ are the results of a geodesic reflection with respect to the connections $\nabla^L$ and $\nabla^{L'}$, we obtain that $\pi^{(v)} : \tilde{q}_0 \mapsto \tilde{q}_1$ since $\pi^{(v)}$ is an affine map. This means that the configuration $\{p_0, q_0, p_1, q_1, p_2, q_2, p_3, q_3\}$ can be closed. $\qquad \square$

**Theorem 4.12** *The local closure conditions for a 3-web manifold $M$ are* **equivalent** *with the following identities for the curvature tensorfield of its Chern connection:*

| Configuration: | Curvature identity: |
|---|---|
| Reidemeister | $R(X, jY)Z = 0$ |
| Moufang | $R(X, jX)Z = R(X, jZ)Z = R(X, jY)X = 0,$ |
| vertical Bol | $R(X, jZ)X = 0,$ |
| horizontal Bol | $R(X, jZ)Z = 0,$ |
| hexagonal | $R(X, jX)X = 0$ |

*for any horizontal vectorfields $X, Y, Z$ on $M$.*

*Proof.* From Theorem 2.10 and Theorem 4.8 we know that the local Reidemeister condition is equivalent to the identity $R(X, jY)Z = 0$ and the local hexagonal condition is characterized by the identity $\sigma\{R(X, jY)Z = 0\}$ for all horizontal vectors $X, Y, Z \in T_p^{(h)}, p \in M$, where $\sigma$ denotes the summation for all permutations of the variables $X, Y, Z$. The last identity is equivalent to the identity $R(X, jX)X = 0$. Theorem 4.10 asserts the equivalence of the vertical local Bol condition and the identity $R(X, jZ)X = 0$. It is clear from the construction of the Chern connection that it is invariant with respect to the interchange of the horizontal and vertical foliations. But by interchanging the vertical and the horizontal foliations the vertical local Bol condition is transformed into the horizontal local Bol condition. It follows that the 3-webs satisfying the horizontal Bol condition can be characterized by the curvature identity $R(X, jZ)X = 0$ for any vertical vectors $X, Z \in T_p^{(v)}, p \in M$. Using $jR = Rj$ (cf. Theorem 1.10) we obtain the curvature condition $jR(X, jZ)X = -R(jZ, j^2X)jX = 0$ for any vertical vectors $X, Z \in T_p^{(v)}, p \in M$. This is equivalent to the curvature identity $R(X, jZ)Z = 0$ for any horizontal vectors $X, Z \in T_p^{(h)}, p \in M$.
From the characterization of 3-webs satisfying both the horizontal and the vertical local Bol conditions we obtain that a 3-web is a Moufang web if and only if the curvature tensorfield is skew-symmetric. $\qquad \square$

## 4.4 Torsionless 3-webs

Let $M$ be a manifold equipped with an $(h, j)$-structure such that the associated Chern connection has vanishing torsion. According to Theorem 4.1 the horizontal, vertical and transversal distributions are integrable, that is the $(h, j)$-structure corresponds to a 3-web.

**Definition 4.13** A 3-web on a manifold $M$ is called *torsionless* if the torsion tensorfield of its Chern connection is identically 0.

In this case of torsionless 3-webs we can find many other distributions, constructed from the horizontal, vertical and transversal distributions, which are integrable. The horizontal, vertical and transversal subspaces of the tangent space $T_pM$ at $p \in M$ determine a 1-parameter family of subspaces in the following way: for a real number $\sigma \in \mathbb{R}$ we consider the subspace

$$T_p^{(\sigma)}M = \{\mathbf{x} - \sigma j\mathbf{x}; \quad \mathbf{x} \in T_p^{(h)}M\}.$$

We have $T_p^{(0)}M = T_p^{(h)}M$ and $T_p^{(1)}M = T_p^{(t)}M$. Putting $T_p^{(\infty)}M = T_p^{(v)}M$ we can see that this family of subspaces is diffeomorphic to the projective line. Any two subspaces in this family are complementary subspaces in $T_pM$. Identifying the family $\{T_p^{(\sigma)}M, \quad \sigma \in \mathbb{R} \cap \{\infty\}\}$ with the projective line $\{\mathbb{R} \cap \{\infty\}\}$ we will refer to the parameter $\sigma \in \{\mathbb{R} \cup \{\infty\}\}$ as the cross-ratio of the subspace $T_p^{(\sigma)}M$ with respect to the horizontal, vertical and transversal subspaces.

**Theorem 4.14** *Let $M$ be a 3-web manifold. A distribution $T_p^{(\sigma)}M$ for a fixed cross-ratio $\sigma \in \{\mathbb{R} \cup \{\infty\}\}$ different from $0, 1$ and $\infty$ is integrable if and only if the torsion tensorfield of the Chern connection vanishes identically. In this case the distribution $T_p^{(\sigma)}M$ is integrable for any cross-ratio $\sigma \in \{\mathbb{R} \cup \{\infty\}\}$.*

*Proof.* Let $X$, $Y$ be horizontal vectorfields and $X - \sigma jX$, $Y - \sigma jY$ sections of the distribution $T^{(\sigma)}M$. Then

$$[X - \sigma jX, Y - \sigma jY] = \sigma[Y, jX] - \sigma[X, jY] + [X, Y] + \sigma^2[jX, jY].$$

Since the horizontal and vertical distributions are integrable we have

$$h[X - \sigma jX, Y - \sigma jY] = \sigma h[Y, jX] - \sigma h[X, jY] + [X, Y]$$

and

$$v[X - \sigma jX, Y - \sigma jY] = \sigma v[Y, jX] - \sigma v[X, jY] + \sigma^2[jX, jY].$$

Hence $[X - \sigma jX, Y - \sigma jY]$ is a section of the distribution $T^{(\sigma)}M$ if and only if

$$-\sigma j(\sigma h[Y, jX] - \sigma h[X, jY] + [X, Y]) = \sigma v[Y, jX] - \sigma v[X, jY] + \sigma^2[jX, jY]$$

or equivalently

$$\sigma v j([jX, j^2 Y] - [jY, j^2 X]) - j[X, Y] = v([Y, jX]) - [X, jY] + \sigma[jX, jY],$$

which can be written as

$$\sigma(\nabla_{jX} jY - (\nabla_{jY} jX - [jX, jY]) = -j(\nabla_X Y - \nabla_Y X - [X, Y]).$$

Since the transversal distribution is integrable we have $T(jX, jY) = -jT(X, Y)$ and hence the distribution $T^{(\sigma)} M$ is integrable for $\sigma \neq 0, 1, \infty$ if and only if the connection is torsionless. But if this condition is fulfilled then we obtain the integrability for any $\sigma \in \mathbb{R}$. □

**Proposition 4.15** *The curvature tensorfield of a torsionless 3-web satisfies*

$$R(X, jY)Z = R(X^\pi, Y^\pi)Z^\pi$$

*for any permutation $\pi$ of the horizontal vectorfields $X, Y, Z$.*

*Proof.* The first Bianchi identity gives for horizontal or vertical vectorfields $X$, $Y$, $Z$ $(\nabla_Z T)(X, Y) = R(Z, jY)X - R(Z, jX)Y$. Since $T(X, Y) = 0$ one has $R(Z, jY)X - R(Z, jX)Y = 0$. It follows that $R(jY, j^2 Z)jX - R(jX, j^2 Z)jY = 0$, or putting $X, Y, Z$ instead of $jX$, $jY$, $jZ$, we obtain $R(Y, jZ)X - R(X, jZ)Y = 0$. Consequently

$$R(X, jY)Z - R(Y, jX)Z = (R(X, jY)Z - R(X, jZ)Y-$$

$$-R(Y, jX)Z + R(Y, jZ)X + R(X, jZ)Y - R(Y, jZ)X = 0,$$

that is we can interchange any two of the tensorfields $X, Y, Z$. □

**Corollary 4.16** *A torsionless 3-web satisfies the local hexagonal condition if and only if it is parallelizable.*

*Proof.* The assertion is a consequence of the previous proposition and of Theorem 4.8. □

# References

1. M. A. AKIVIS, *Three-Webs of multidimensional surfaces* (Russian), Trudy Geom. Sem. **2** (1969), 7–31.
2. M. A. AKIVIS, *Differential geometry of webs* (Russian), Problems in Geometry, Itogi Nauki i Techniki **15** (1983), 187–213.

3. M. A. AKIVIS & A. M. SHELEKHOV, *The computation of the curvature and torsion tensors of a multidimensional three-web and of the associator of the local quasigroup connected with it* (Russian), Sibirsk. Math. Zh. **12** (1971), 953–960.

4. M. A. AKIVIS & A. M. SHELEKHOV, *Foundation of the theory of webs* (Russian), Kalinin University, Kalinin, 1981.

5. M. A. AKIVIS & A. M. SHELEKHOV, *Geometry and Algebra of Multidimensional Three-Webs*, Kluwer Academic Publishers, Dordrecht –Boston – London (1992).

6. A. BARLOTTI & K. STRAMBACH, *The geometry of binary systems*, Advances in Math. **49** (1983), 1–105.

7. W. BLASCHKE & G. BOL, *Geometry der Gewebe*, Springer-Verlag, Berlin, 1938.

8. W. BLASCHKE, *Einführung in die Geometrie der Waben*, Birkhäuser-Verlag, Basel-Stuttgart, 1955.

9. S. S. CHERN, *Eine Invariantentheorie der Dreigewebe aus r-dimensionalen Mannigfaltigkeiten in* $\mathbb{R}_{2r}$, Abh. Math. Sem. Univ. Hamburg **11** (1936), 335–358.

10. V. I. FEDOROVA, *A condition defining multidimensional Bol's three-webs* (Russian), Sibirsk. Mat. Zh. **19** (1978), 922–928.

11. M. FUNK & P. T. NAGY, *On collineation groups generated by Bol reflections*, Journal of Geometry **48** (1993), 63–78.

12. K. H. HOFMANN & K. STRAMBACH, *Topological and analytical loops*, Chap. VIII in *Quasigroups and Loops: Theory and Applications* (Ed.: O. Chein, H. O. Pflugfelder and J. D. H. Smith), Sigma Series in Pure Math. 8, Heldermann-Verlag, Berlin (1990), 205–262.

13. F. S. KERDMAN, *Moufang loops in the large* (Russian). Algebra i Logika **18** (1979), 523-555.

14. S. KOBAYASHI & K. NOMIZU, *Foundations of Differential Geometry*, Vol. I., Interscience, New York, 1963.

15. S. KOBAYASHI & K. NOMIZU, *Foundations of Differential Geometry*, Vol. II., Interscience, New York, 1969.

16. E. N. KUZ'MIN, *The connection between Malcev algebras and analytic Moufang loops* (Russian), Algebra i Logika **10** (1971), 3–22.

17. A. I. MALCEV, *Analytic loops* (Russian), Mat. Sb. **36** (1955), 569–576.

18. P. O. MIHEEV & L. V. SABININ, *The Theory of Smooth Bol Loops*, Friendship of Nations University, Moscow, 1985.

19. P. O. MIHEEV & L. V. SABININ, *Quasigroups and differential geometry*, Chap. XII in *Quasigroups and Loops: Theory and Applications* (Ed.: O. Chein, H. O. Pflugfelder and J. D. H. Smith), Sigma Series in Pure Math.

8, Heldermann-Verlag, Berlin (1990), 357–430.

20. P. T. NAGY, *Invariant tensorfields and the canonical connection of a 3-web* Aequationes Math. **35** (1988), 31–44.

21. P. T. NAGY, *On complete group 3-webs and 3-nets*, Arch. Math. **53** (1989), 411–413.

22. P. T. NAGY, *3-nets with maximal family of two-dimensional subnets*, Abh. Math. Sem. Hamburg **61** (1991), 203–211.

23. P. T. NAGY, *Extension of local loop isomorphisms*, Monatshefte Math. Wien **112** (1991), 221–225.

24. P. T. NAGY, *Moufang loops and Malcev algebras*, Seminar Sophus Lie (Darmstadt) **3** (1992), 65–68.

25. P. T. NAGY & K. STRAMBACH, *Loops, their cores and symmetric spaces*, Israel J. Math. **105** (1998), 285–322.

26. P. T. NAGY & K. STRAMBACH, *Sharply transitive sections in Lie groups: a Lie theory of smooth loops*, manuscript.

27. J. A. WOLF, *Spaces of Constant Curvature*, McGraw-Hill, New York, 1967.

# CONFORMAL FLOWS ON $\mathbb{C}_{,0}$ AND HEXAGONAL 3-WEBS

MICHEL BELLIART, ISABELLE LIOUSSE AND FRANK LORAY

*Laboratoire de Géométrie, Analyse et Topologie, C.N.R.S. URA D751, U.F.R. de Mathématiques, Université Lille I, 59655 Villeneuve d'Ascq Cedex, France*

*E-mail : belliart,liousse,loray@gat.univ-lille1.fr*

One proves a rigidity Lemma for Lie pseudo-groups generated by conformal flows. One deduces that when one lifts any non-solvable dynamics of $\mathrm{Aut}(\mathbb{C}_{,0})$ to the tangent bundle over each Nakai sector, every orbit is dense. In particular, one gives a proof of the density on the unitary bundle using hexagonal 3-webs; this provides a new and elementary proof of a rigidity theorem due to A.A. Shcherbakov, following an idea of I. Nakai.

This work is a continuation of [1]; the language is the same.

## 1 The rigidity lemma

Consider two germs of holomorphic vector fields $X$ and $Y$ at $0 \in \mathbb{C}$, with transversal real parts. Choose an open neighborhood $U$ of $0$ on which $X$ and $Y$ are well defined and denote by $\mathcal{G}$ the pseudo-group generated, on this open set, by the flows $\varphi_X^t$ and $\varphi_Y^t$, $t \in \mathbb{R}$. These are the flows of the real parts of $X$ and $Y$ or, equivalently, the complex flows of $X$ and $Y$ restricted to real time. Hence, the flows considered are conformal and transversal and the pseudo-group $\mathcal{G}$ is transitive on any restriction of $U$. We show that, in general, $\mathcal{G}$ is also transitive on the tangent bundle minus the zero section. Precisely:

**Lemma.** *With the above notations, if the subgroup of $\mathbb{C}^*$ defined by:*

$$\mathcal{G}_0 = \{\varphi'(0) \in \mathbb{C}^* \; ; \; \varphi \in \mathcal{G}, \; \varphi(0) = 0\}$$

*is strict, i.e. $\mathcal{G}_0 \neq \mathbb{C}^*$, then the real Lie algebra $\mathcal{A}$ generated by the (real parts of the) vector fields $X$ and $Y$ is at most 3-dimensional.*

**Examples.**

1. If $\mathcal{G}$ is the pseudo-group of translations on $U$, then $\mathcal{A}$ is 2-dimensional and $\mathcal{G}_0 = \{1\}$.

2. If $\mathcal{G}$ is the restriction to $U \subset \mathbb{D}$ of the group of isometries of the disc $\mathbb{D}$ generated, for instance, by two horocyclic flows then $\mathcal{G}_0 \subset \mathbb{S}^1$ by Schwarz's Lemma.

3. If $\mathcal{A}$ is 3-dimensional generated by the vector fields $\frac{\partial}{\partial z}$ and $\lambda z \frac{\partial}{\partial z}$, $\lambda \in \mathbb{C}^*$, at a generic point of the plane $\mathbb{C}$, then $\mathcal{G}$ is a subgroup of the affine group and $\mathcal{G}_0 \subset \{e^{\lambda t} \; ; \; t \in \mathbb{R}\}$.

4. If $\mathcal{G}$ is the group of those rotations of the 2-sphere $\mathbb{S}^2$ generated, at a generic point of the Riemann sphere $\bar{\mathbb{C}}$ , by the rotations around two different axes, then again $\mathcal{G}_0 \subset \mathbb{S}^1$.

We will see that these examples exhaust all possibilities.

## 2  Applications

Denote $\mathrm{Aut}(\mathbb{C}_0)$ the group for the composition law of germs of invertible transformations of $\mathbb{C}$ at 0:

$$\mathrm{Aut}(\mathbb{C}_0) = \{f(z) = az + \ldots ; f \in \mathbb{C}\{z\} \text{ and } a \in \mathbb{C}^*\}$$

and define the dynamics of a finitely generated subgroup $\widehat{G}$ of $\mathrm{Aut}(\mathbb{C}_0)$ as follows. Consider a system of generators $f_1, \ldots, f_p \in \mathrm{Aut}(\mathbb{C}_0)$ of $\widehat{G}$ and a neighborhood $U$ of $0 \in \mathbb{C}$ on which $f_i$ and its inverse $f_{i+p} = f_i^{\circ(-1)}$ are both well defined for $i = 1, \ldots, p$. We are interested in the dynamics of the pseudo-group $G$ generated by $f_i : f_i^{-1}(U) \mapsto U; \; i = 1, \ldots, 2p$. Say that two points $z, z' \in U$ are in the same *orbit* under $G$ if there exists a sequence of points $z = z_1, z_2, \ldots, z_n, z_{n+1} = z'$ of $U$ such that $z_{k+1}$ is the image of $z_k$ by one of the $f_i$ for each $k = 1, \ldots, n$. Applying the rigidity Lemma to the pseudo-group $G$, one gets:

**Theorem.** *If $\widehat{G}$ is non solvable, then the set:*

$$\{(z_0, g'(z_0)) \; ; \; z_0 \in U^* = U \setminus \{0\}, \; g \in G \text{ and } g(z_0) = z_0\}$$

*is dense in $U^* \times \mathbb{C}^*$.*

The proof of the density in $U$ of the set of points $z_0$ fixed by non trivial elements of $G$ has been outlined in 1986 by A.A. Shcherbakov. The calculations involved are surely not straightforward (a recent complete calculation can be found in [13]). In 1995, X. Gomez-Mont and B. Wirtz proved the preceding Theorem under additional assumptions (density of the linear part of $\widehat{G}$). In 1996, J. Rebelo outlined a proof of the existence (and in fact of the density) of attractive fixed points with no other hypothesis than the non-solvability of $\widehat{G}$. His approach suggested to us another proof with no calculation (see [1]) using Nakai's construction of flows. At that time, we had a weaker rigidity Lemma;

the same proof with the presently stronger rigidity Lemma leads to the theorem above. A new proof of Rebelo's result has recently been announced by B.Wirtz; it looks closer to the calculatory approach of Shcherbakov.

*Idea of the proof.* In [12], I. Nakai constructs, in a neighborhood of any point $z_0 \in U^*$, conformal flows $\varphi_X^t, \varphi_Y^t, t \in \mathbb{R}$, which belong to the closure $\overline{G}$ of $G$ for the topology of uniform convergence. *A posteriori*, one can observe that the holomorphic vector fields $X$ and $Y$ are, up to multiplication by a scalar, nothing else but the *iterative logarithms* (see [5]) of the tangent to identity elements of $\widehat{G}$ used in Nakai's construction, from which one immediately deduces that the real Lie algebra $\mathcal{A}$ generated by $X$ and $Y$ is infinite-dimensional. From our rigidity Lemma, for all $a \in \mathbb{C}^*$, there exists an element $g$ of the pseudo-group $\mathcal{G} \subset \overline{G}$ generated by the flows $X$ and $Y$ such that $g(z_0) = z_0$ and $g'(z_0) = a$. The approximants $g_n \in G$ of $g$ must fix a point arbitrary close to $z_0$ (Rouché-Hurwitz) with a derivative arbitrary close to $a$. □

A similar use of our rigidity Lemma permits to improve the:

**Theorem (Shcherbakov-Nakai).** *If $\widehat{G}$ is non solvable, then:*

- *either each orbit of the pseudo-group $G$ is dense on $V := U\setminus\{0\}$,*

- *or there exists an integer $\nu \in \mathbb{N}^*$ and a conformal coordinate in which the pseudo-group $G$ leaves the analytic set $\Sigma = \{z \in U; z^\nu \in \mathbb{R}\}$ invariant and then each orbit is dense in each sector $V \subset U\setminus\Sigma$ it reaches.*

In fact, A.A. Shcherbakov's result asserts that there exists a finite collection of disjoint open sets whose union is dense in $U$, and in which each $G$-orbit is dense. Nakai's contribution consisted not only in giving a new proof based on his construction of conformal flows in $\overline{G}$, but also in making this collection of open sets more precise. Moreover, this new proof shows the genericity of the first alternative. Denote $T^*V$ the tangent bundle of $V$ deprived of the zero section, $T^*V \simeq V \times \mathbb{C}^*$. One has:

**Corollary.** *If $\widehat{G}$ is non solvable, and if we still denote by $G$ the same pseudo-group acting on the tangent bundle, then each $G$-orbit is dense in each sectorial sub-bundle $T^*V$ of $T^*U$ it reaches.*

I. Nakai has recently and independently proved the sectorial density on the unitary bundle which is a particular case of our Corollary. Our motivations are essentially the same, namely to give an elementary proof of:

**Theorem (Shcherbakov).** *If $\widehat{G}$ is non solvable and if there exists an orientation preserving homeomorphism $h : U \to U' \subset \mathbb{C}$ ; $0 \mapsto 0$ conjugating the*

*dynamics of G with those of another pseudo-group of conformal transforma-*
*tions, then h is in fact conformal.*

Using his construction of flows, I. Nakai proves immediately the real dif-
ferentiability of $h$. Nevertheless, the final argument given in [12] to show the
conformality of $h$ remained uncomplete. The idea is to look at the ellipse
bundle defined over $U$ by the pull-back of the conformal structure on $U'$ via
$h$: by construction, the pseudo-group $G$ must preserve it. Now, using our
Fixed Points Theorem or, equivalently, the sectorial density on the unitary
bundle recently proved by Nakai, one deduces that this invariant ellipse field
has to be a circle field (everywhere, ellipses are turning under the action of
$G$), and then $h$ must be conformal. Actually, Shcherbakov's Theorem asserts
that there does not exist any deformation of the complex structure of such a
dynamical system.

## 3   Motivations

Since Poincaré, qualitative problems concerning continuous dynamical systems of two variables can be reduced, via holonomy mappings, to questions
of the same nature for discrete dynamical systems of one variable. When
studying singularities of holomorphic foliations in complex dimension 2, one
is interested in pseudo-groups of one complex variable near a common fixed
point. These dynamics describe a part of the transverse structure of those
singularities (see for instance [4]) and are in general non solvable (see [9]). The
existence of attractive fixed points for the transverse structure corresponds
to the existence of limit cycles for the foliation. Otherwise, Shcherbakov's
theorem allowed to prove the:

**Theorem (Loray).** *Let $\mathcal{F}$ be a germ of singular holomorphic foliation at the*
*origin of $\mathbb{C}^2$ defined by a Pfaffian equation $\omega = 0$ where:*

$$\omega = d(y^2 - x^3) + \text{ terms of order higher than 3 in } x \text{ and } y.$$

*If $\mathcal{F}$ is mapped by a germ of an orientation preserving homeomorphism to*
*another germ of holomorphic foliation $\mathcal{F}'$, then $\mathcal{F}$ and $\mathcal{F}'$ are also conjugated*
*by a germ of biholomorphism.*

Another use of the existence of attractive fixed points can be found in [11].

## 4   Proof of the rigidity in the particular case $G_0 = \{1\}$

More generally, we have the:

**Proposition.** *Let $X_1, \ldots, X_n$ be germs of real $C^\infty$ vector fields at the origin of $\mathbb{R}^m$ such that the pseudo-group $\mathcal{G}$ generated by their flows is transitive in a neighborhood of the origin. If the subgroup of $Gl(m, \mathbb{R})$ defined by:*

$$\mathcal{G}_0 := \{T_0 g \; ; \; g \in \mathcal{G}, \; g(0) = 0\}$$

*is trivial, i.e. reduces to the identity, then the Lie algebra $\mathcal{A}$ generated by the vector fields $X_1, \ldots, X_n$ is at most $m$-dimensional.*

This result is a local version of [8]; it is certainly well known by this author.

*Proof.* The hypothesis $\mathcal{G}_0 = \{id\}$ exactly means that any germ $Y$ of vector field which is left invariant by $\mathcal{G}$ is constructed by propagating the vector $Y(0)$ by $\mathcal{G}$. In particular, such invariant vector fields form an $m$-dimensional vector space which is identified with the fiber $T_0 \mathbb{R}^m$. Choose $m$ generators $Y_1, \ldots, Y_m$ and denote $\mathcal{G}'$ the transitive pseudo-group generated by their flows. The vector fields $Y_1, \ldots, Y_m$ commute with $X_1, \ldots, X_n$ and then with the Lie algebra $\mathcal{A}$ they generate (Jacobi identity). Reversing the preceding argument, each element $X \in \mathcal{A}$ is constructed by propagating $X(0)$ by the pseudo-group $\mathcal{G}'$ and so is given by $X(0) \in T_0 \mathbb{R}^m$. □

## 5  Proof of the rigidity in the particular case $\mathcal{G}_0 \subset \mathbb{S}^1$

We reproduce here the proof of [1].

Over $\mathbb{R}$, the hypothesis asserts that $\mathcal{G}_0$ is contained in the subgroup of isometries of $Gl(2, \mathbb{R})$. So, by propagating the usual euclidian metric defined on the tangent plane $T_0 \mathbb{R}^2$ by $\mathcal{G}$, we construct a riemannian metric which is $\mathcal{G}$-invariant in a neighborhood of the origin. Then $\mathcal{G}$ is a transitive pseudo-group of isometries, which means that the metric has constant curvature and is hence equivalent to one of the three geometries (hyperbolic, euclidian or spheric), and moreover that the elements of $\mathcal{G}$ belong to a well-known 3-dimensional Lie group.

## 6  Hexagonal 3-webs

A *3-web* on an open set of $\mathbb{R}^2$ consists of 3 pairwise transversal regular foliations $(\mathcal{F}_1, \mathcal{F}_2, \mathcal{F}_3)$ defined, for instance, by pfaffian equations $\omega_i = 0$, $i = 1, 2, 3$, where the $\omega_i$ are non vanishing differentiable 1-forms. In particular, a *linear 3-web* is given by 3 families of parallel lines and the *equilateral 3-web* is the linear 3-web of slopes 0, $\sqrt{3}$ and $-\sqrt{3}$. All linear 3-webs are equivalent by linear change of coordinates.

If all 1-webs and 2-webs[a] are locally equivalent, this is not so for 3-webs. Call a 3-web *linearizable* if it is equivalent, by a change of coordinates, to *the* linear 3-web.

In 1927, Thomsen proposed the following characterization of linearizability of 3-webs. We say that a 3-web is *hexagonal at a point P* if, starting from any point $Q$ near $P$ on any of the 3 leaves passing through $P$, and turning around $P$, moving successively along the 3 foliations as a spider builds its web, the picture closes into an hexagon:

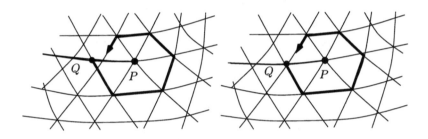

Figure 1.

We say that a 3-web is *hexagonal* if it is at every point. The linear 3-web is of course hexagonal and a 3-web which is not is certainly non linearizable.

**Example.** The 3-web defined by the 3 pencils of projective lines passing through 3 distinct points $P_1$, $P_2$ and $P_3$ on the projective plane $\mathbb{RP}^2$ is hexagonal on the dense open set where it is regular (i.e. outside the tangencies).

Indeed, in an affine chart containing the 3 points, hexagonality is nothing but Pappus's theorem.

*Remark.* Hexagonality means that each sufficiently small triangle inscribed in the web is the element of a tiling also inscribed in the web:

One can always linearize the two first foliations of a 3-web into the horizontal and vertical lines. The local study of the 3-webs is then reduced to the local study of a foliation (the third one) or, equivalently, of a differential equation $\frac{dy}{dx} = f(x,y)$, $f(0,0) \neq 0$, up to changes of coordinates of the type[b] $(x,y) \mapsto (\varphi(x), \psi(y))$. From this point of view, E. Cartan showed that the Lie pseudo-group leaving invariant a 3-web is 0-, 1- or 3-dimensional. The first

---

[a] A 1-web is a foliation and a 2-web is made of two transversal foliations.
[b] Precisely the changes of coordinates which leave the linear 2-web invariant.

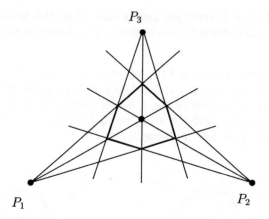

$P_3$

$P_1$

$P_2$

Figure 2.

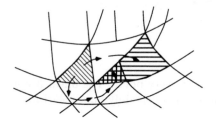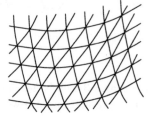

Figure 3.

case is the generic case. In the second case, the differential equation can be brought to the form $\frac{dy}{dx} = f(x + y)$ and in the third case, to the form $\frac{dy}{dx} = 1$: this is the linearizable case. For the equilateral model of the linear 3-webs, Cartan's 3-parameters group is the subgroup of the affine group of transformations on the real plane $\mathbb{R}^2$ generated by the translations, the dilatations and the periodic rotations of order 3.

Let us describe now the differential geometric approach proposed by Blaschke. A 3-web can always be defined by differential 1-forms $(\omega_1, \omega_2, \omega_3)$ satisfying:

$$\omega_1 + \omega_2 + \omega_3 = 0.$$

Under this condition, we define without ambiguity:

$$\Omega := \omega_1 \wedge \omega_2 = \omega_2 \wedge \omega_3 = \omega_3 \wedge \omega_1.$$

The 2-form $\Omega$ is regular. We define functions $h_1$, $h_2$ and $h_3$ by:

$$d\omega_i = h_i.\Omega, \quad i = 1,2,3.$$

and unambiguously the 1-form:

$$\theta := h_3\omega_2 - h_2\omega_3 = h_1\omega_3 - h_3\omega_1 = h_2\omega_1 - h_1\omega_2.$$

Now, it is easily checked that $\theta$ is the unique differential 1-form satisfying:

$$d\omega_i = \theta \wedge \omega_i, \quad i = 1,2,3,$$

and that the 2-form $d\theta$ does not depend on the 1-forms $\omega_i$ chosen to define the 3-web: $d\theta$ is the *curvature 2-form* of the 3-web $(\mathcal{F}_1, \mathcal{F}_2, \mathcal{F}_3)$.

**Theorem.** *Let $(\mathcal{F}_1, \mathcal{F}_2, \mathcal{F}_3)$ be a germ of a 3-web at the origin of $\mathbb{R}^2$ defined by 1-forms $\omega_i$, $i = 1,2,3$, satisfying $\omega_1 + \omega_2 + \omega_3 = 0$ and let $\theta$ be the unique differential form satisfying $d\omega_i = \theta \wedge \omega_i$, $i = 1,2,3$, as above. Then, the following are equivalent:*

(i) *the 3-web is linearizable,*

(ii) *the 3-web is hexagonal,*

(iii) *the 3-web is left invariant by a transitive pseudo-group $G$,*

(iv) *the 3-web has zero curvature: $d\theta \equiv 0$.*

*Proof. $(i) \Rightarrow (ii)$ is trivial.*

To prove $(ii) \Rightarrow (iii)$, we define in the neighborhood of each leaf $L$ of each of the 3 foliations an involutive transformation $\sigma_L$, $\sigma_L \circ \sigma_L = identity$, in the following way:

By construction, $\sigma_L$ leaves the leaf $L$ point-wise invariant and exchanges the two transversal foliations: this is *the* reflection associated with $L$ within the 3-web.

Now the following picture, together with a standart density argument on the dyadic numbers, shows that $\sigma_L$ leaves also the tangent foliation (whence the 3-web) invariant assuming the web is hexagonal.

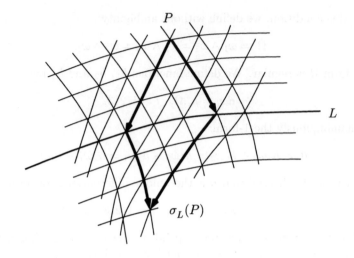

Figure 4.

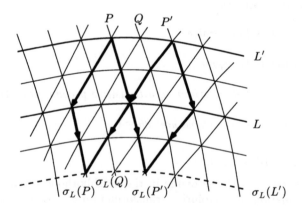

Figure 5.

Note, although it won't be used here, that the 3-web is hexagonal at a point $P$ if and only if the reflections relative to the 3 leaves passing through

$P$ are such that $\sigma_1 \circ \sigma_2 \circ \sigma_3$ is an involution[c].

A *posteriori*, the reflections $\sigma_L$ play the role of the orthogonal symmetries of the equilateral 3-web relatively to its lines ; the transformations $\sigma_L \circ \sigma_{L'}$ correspond then to the order 3 rotations, when $L$ and $L'$ intersect, and to the translations, when they belong to the same foliation. One could prove directly, again by a density argument on the dyadic numbers, that the set of transformations $\sigma_L \circ \sigma_{L'}$, when $L$ and $L'$ span the set of the leaves of one of the 3 foliations, is a one parameter group. Nevertheless, one easily sees that the pseudo-group $\mathcal{G}$ generated by those 3 "translation flows" is transitive, which is sufficient for our purpose.

We will need $(iii) \Rightarrow (ii)$. Given any sufficiently small triangle, we construct a local pavement by pushing around by $\mathcal{G}$ the given triangle and its reflection with respect to one of its sides: this implies hexagonality.

In order to show $(iii) \Rightarrow (iv)$ we recall a remark of [6]: "a 3-web defines a complex structure". Indeed, at every point $P$, there exists on the tangent plane $T_P \mathbb{R}^2$ a unique linear transformation $J_P$ satisfying $J_P^2 + J_P + I = 0$: this is the order 3 linear transformation which cyclically permutes the 3 directions defined by the web. We thus obtain a $J$-structure on our surface; according to a well-known Integrability Theorem, one can assume that, up to change of coordinates, this $J$-structure is the standard one defined by the usual order 3 rotations. In other words, the 3 given direction fields endow the unitary bundle with an integrable conformal structure which is $\mathcal{G}$-invariant. Suppose the 2-form $d\theta \neq 0$ at a point $P$. Then $d\theta \neq 0$ everywhere because it is $\mathcal{G}$-invariant and defines a volume element. Since we are able to measure angles and volumes, we have defined an invariant metric which brings us back to §4: $\mathcal{G}$ is a locally transitive pseudo-group for one of the 3 homogeneous metrics. Moreover, since the invariant 3-web allows one to construct tilings by equilateral triangles, we are in the euclidian case. Now, the 3-web is constructed by propagating the 3 directions at the point $P$ by translations, so it has to be linear.

Finally, let us prove $(iv) \Rightarrow (i)$. The hypothesis $d\theta \equiv 0$ implies that one can write $\theta = \frac{dF}{F}$ where $F$ is a nowhere-vanishing function. Then $F$ is a common integrating factor for the 3 1-forms $\omega_i$, i.e. $d(\frac{\omega_i}{F}) \equiv 0$, and by integrating the identity $\frac{\omega_1}{F} + \frac{\omega_2}{F} + \frac{\omega_3}{F} = 0$, one obtains $H_1 + H_2 + H_3 = 0$

---

[c]This also means that the group generated by those 3 reflections is metabelian and is to be thought of as a local version of the following famous corollary of Pascal's Theorem: "the product of 3 involutions of $PGL(2, \mathbb{R})$ is again an involution iff these involutions belong to the same metabelian group".

where $H_1$, $H_2$ and $H_3$ are first integrals of the respective foliations $\mathcal{F}_1$, $\mathcal{F}_2$ and $\mathcal{F}_3$. In the coordinate system $(H_1, H_2)$, the 3-web is linear. $\qquad\square$

# 7  Proof of the rigidity in the particular case $\mathcal{G}_0 \subset \mathbb{R}$

By hypothesis, each direction (real line) of the tangent plane $T_0\mathbb{C}$ propagates by the pseudo-group $\mathcal{G}$ into an invariant foliation. If we choose 3 directions, we then obtain an invariant 3-web which, by the §5 Theorem, is hexagonal. In a good system of real coordinates, this web is equilateral and the pseudo-group contains only translations and (real) homotheties. The Lie algebra $\mathcal{A}$ is then 3-dimensional.

# 8  Proof of the rigidity in the general case

The group $\mathcal{G}_0$ is a connected subgroup of $\mathbb{C}^*$. Indeed, each transformation $\varphi_X^{r_1} \circ \varphi_Y^{s_1} \circ \ldots \circ \varphi_X^{r_n} \circ \varphi_Y^{s_n}$ can[d] be isotopied to the identity in the isotropy pseudo-group:

$$\mathcal{G}^{\{0\}} := \{g \in \mathcal{G}; \ g(0) = 0\}$$

by pulling continuously the integration times $r_i, s_j \in \mathbb{R}$ back to 0. Now, suppose that $G$ is not all of $\mathbb{C}^*$. Then, because $\mathcal{G}$ can't be trivial (see §4), one easily deduces that $\mathcal{G}_0$ is a real 1 dimensional submanifold of $\mathbb{C}^*$. Let the pseudo-group $\mathcal{G}$ act on the tangent bundle $TU$. In the trivialization:

$$U \times \mathbb{C} \longrightarrow TU; (z,t) \longmapsto (z, t\frac{\partial}{\partial z}),$$

the pseudo-group action $z \mapsto g(z)$ is lifted to $(z,t) \mapsto (g(z), g'(z).t)$. All the orbits but the zero section are real 3-dimensional submanifolds transverse to the fibers because their restriction to each fiber is a coset for the $\mathcal{G}_0$ action on $\mathbb{C}^*$. Consider the 2-dimensional complex manifold $M = T^*U := U \times \mathbb{C}^*$. If an element $g \in \mathcal{G}$ fixes some point $P = (z,t) \in M$, then its differential at $P$ is:

$$Dg(z,t) = \begin{pmatrix} g'(z) & 0 \\ g''(z).t & g'(z) \end{pmatrix} = \begin{pmatrix} 1 & 0 \\ g''(z).t & 1 \end{pmatrix}.$$

On the other hand, this differential fixes a non vertical complex direction in $T_PM$; indeed, $g$ leaves the 3-dimensional real subspace tangent to the orbit $\mathcal{G}$ invariant and then the unique complex line contained in it also (it can't be

---

[d]Here, we need $U$ to be simply connected.

vertical because the 3-dimensional orbits must generate the horizontal transitivity). In particular, the coefficient $g''(z).t$ has to be zero: the differential of an element $g \in \mathcal{G}$ at a point $P \in M$ fixed by $g$ must be the identity.

By restricting the action $\mathcal{G}$ to one of the 3-dimensional orbits, proposition of §4 applies with $n = 2$ and $m = 3$: $\mathcal{A}$ is then a 3-dimensional Lie algebra.

Let us check that the list given in §0 is complete. If the real Lie algebra $\mathcal{A}$ is at most 3-dimensional, the same holds true for the complex Lie algebra generated by $X$ and $Y$. In this case it is well known that up to an analytic change of coordinates, this algebra is a subalgebra of the Lie algebra of $PSL(2, \mathbb{C})$. We can then suppose that $X$ and $Y$ are both of the form $(az^2 + bz + c)\frac{\partial}{\partial z}$, $a, b, c \in \mathbb{C}$. Sections 3 and 6 completely deal with the situations $\dim_{\mathbb{R}} \mathcal{A} = 2$ and $\mathcal{G}_0 = \mathbb{R}^+$. Roughly speaking, in the general case, the Lie algebra $A$ is 3-dimensional, two dimensions are used to move transitively in the neighborhood of $0 \in \mathbb{C}$ and the remaining dimension acts on the tangent plane $\mathcal{G}_0 = \{e^{t\lambda}; t \in \mathbb{R}\}$, $\lambda \in \mathbb{C}^*$, $\lambda^2 \notin \mathbb{R}$. Then, the Lie algebra $\mathcal{A}$ is also generated over $\mathbb{R}$ by a regular vector field $X = (az^2 + bz + c)\frac{\partial}{\partial z}$, $c \neq 0$, and a vector field of the form $(dz^2 + \lambda z)\frac{\partial}{\partial z}$ which can be supposed linear by a Moebius change of coordinates : $Y = (\lambda z)\frac{\partial}{\partial z}$. We now use the hypothesis that the real Lie algebra generated by $X$ and $Y$ is 3-dimensional. Suppose that $a \neq 0$ in the expression of $X$, i.e. up to a linear change of coordinates, $a = 1$. Then, the 4 vector fields:

$$X = (z^2 + bz + c)\frac{\partial}{\partial z}$$

$$Y = (\lambda z)\frac{\partial}{\partial z}$$

$$Z := [X, Y] = (-\lambda z^2 + \lambda c)\frac{\partial}{\partial z}$$

$$\text{and } [Z, Y] = (\lambda^2 z^2 + \lambda^2 c)\frac{\partial}{\partial z}$$

are independent over $\mathbb{R}$: in fact, $a = 0$, $X = (bz + c)\frac{\partial}{\partial z}$ and $\mathcal{G}$ is an affine transformation pseudo-group. $\square$

## Credits

We wish to thank Isao Nakai for valuable discussions, and hope to have emphasize some of his ideas.

104

**References**

1. M. BELLIART, I. LIOUSSE & F. LORAY, *Sur l'existence de points fixes attractifs pour les sous-groupes de Aut($\mathbb{C},0$)*, C. R. Acad. Sci. Paris **324**, Série I (1997), 443–446

2. W. BLASCHKE & G. BOL, *Geometrie der Gewebe*, Springer, Berlin, 1938

3. E. CARTAN, *Les sous-groupes des groupes continus de transformations*, Œuvres complètes, vol. 3, 78–83.

4. D. CERVEAU & R.MOUSSU, *Groupes d'automorphismes de $(\mathbb{C},0)$ et équations différentielles $ydy + \ldots = 0$*, Bull. Soc. Math. France **116** (1988), 459–488

5. J. ECALLE, *Théorie itérative : introduction à la théorie des invariants holomorphes*, J. Math. Pures et Appl. **54** (1975), 183–258

6. E. GHYS, *Flots transversalement affines et tissus feuilletés*, Mémoires de la S.M.F. **46** (1991), 123-150

7. X. GOMEZ-MONT & B. WIRTZ, *On fixed points of conformal pseudogroups*, Bol. Soc. Bras. Mat. **262** (1995), 201-209

8. S. KOBAYASHI, *Le groupe de transformations qui laissent invariant un parallélisme*, Colloque de topologie de Strasbourg, 1954–55.

9. L. LE FLOCH, *Théorème de rigidité pour une famille à un paramètre d'équations différentielles holomorphes*, C. R. Acad. Sci. Paris **319** (1994), 1197–1200

10. F. LORAY, *Rigidité topologique pour des singularités de feuilletages holomorphes*, Ecuaciones diferenciales y singularidades, J.M. Fernandez, Universidad de Valladolid (Spain), (1997), 213–234

11. J.-F. MATTEI & E. SALEM, *Complete systems of topological and analytical invariants for a generic foliation of $(\mathbb{C}^2,0)$*, Math. Res. Lett. **4** (1997), no. 1, 131–141

12. I. NAKAI, *Separatrices for non solvable dynamics on $\mathbb{C},0$*, Ann. Inst. Fourier **44** (1994), no. 2, 569–599

13. L. ORTIZ, E. ROSALES & A.A. SHCHERBAKOV, *A remark on a countable set of limit cycles for the equation $dw/dz = P_n(z,w)/Q_n(z,w)$*, in preparation

14. J. REBELO, *oral communication*

15. A.A. SHCHERBAKOV, *On the density of an orbit of a pseudogroup of conformal mappings and a generalization of the Hudai-Verenov theorem*, Vestnik Movskovskogo Universiteta **31** (1982), no. 4, 10–15

16. A.A. SHCHERBAKOV, *Topological and analytic conjugation of noncommutative groups of germs of conformal mappings*, Trudy Sem. Petrovskogo **10** (1984), 170–196, 238–239;

17. A.A. SHCHERBAKOV, *On complex limit cycles of the equation $dw/dz =$ $P_n/Q_n$*, Russian Math. Surveys **41** (1986), no. 1, 243–244

# RIGIDITY OF WEBS

JEAN-PAUL DUFOUR

*Département de Mathématiques - Université de Montpellier II*
*Place Eugène Bataillon, 34095 Montpellier Cedex 05,*
*dufourj@darboux.math.univ-montp2.fr*

## 1  Webs

*A codimension c d-web on a manifold V is a family* $(\mathcal{F}_1, ..., \mathcal{F}_d)$ *of d foliations on V, each of codimension c, and in general position.*

It seems that the first webs which attracted attention of mathematicians were duals of projective curves (see Henaut's paper in this volume). In fact, this volume shows that we fall on webs in many domains: implicit differential equations, study of connections, P.D.E., integrable systems... I encountered my first web when I tried, in the 70ths, to understand families of plane curves: for these families there appear generically cuspidal envelops; there are three different curves which pass at each point in the interior of the cusp, so these curves form a 3-web (of codimension 1). The invariants of this web (such as the curvature) give differential invariants for these families of curves which prevent, for example, differentiable stability. However R. Thom had raised the problem of the possible *topological* stability. It appeared later that the topological stability is equivalent to the differentiable stability. More precisely: an homeomorphism between two such webs is automatically a diffeomorphism. In fact there are very few homeomorphisms which preserve simultaneously the three foliations: they form a finite dimensional Lie group and all of them are differentiable. This is what we call the *rigidity* of these webs. In what follows I will give a proof of this rigidity for 3-webs of codimension $n$ in a $2n$-dimensional manifold. A similar proof shows the rigidity of $(k + 1)$-webs of codimension $n$ in a $kn$-dimensional manifold.

In this paper our webs $(\mathcal{F}_1, ..., \mathcal{F}_d)$ are always smooth (i.e. $C^\infty$) or analytic. If the codimension is $c$ and if the dimension of $V$ is not less than $dc$ then, near every point of $V$, there are local coordinates

$$x_1^1, \ldots, x_c^1, x_1^2, \ldots, x_c^2, \ldots \ldots, x_1^d, \ldots, x_c^d, y_1, \ldots$$

such that $\mathcal{F}_i$ is given by $x_1^i = \text{constant},\ldots,x_c^i = \text{constant}$, for $i = 1\cdots d$. So we see that this web is not rigid: there are non-smooth local homeomorphisms (products of homeomorphisms in the $(x_1^i, \ldots, x_c^i)$ spaces) which preserve simultaneously each foliation. The rigidity phenomenon appears when $cd$ is strictly greater than the dimension of $V$. To obtain our rigidity theorem (theorem 4), we will first give a "normal form" for 3-webs. Such normal forms exist for more general webs, but in the case of 3-webs it can also be seen as a generalization of the classical Campbell-Hausdorff formula for Lie groups: we will see that 3-webs are deeply related to "smooth loops" which are generalizations of Lie groups. This normal form is unique, up to linear isomorphisms, and it contains all local invariants of 3-webs.

## 2 Smooth loops and 3-webs

In the sequel we will consider a 3-web $(\mathcal{F}_1, \mathcal{F}_2, \mathcal{F}_3)$ of codimension $n$ in a $2n$-manifold $V$.

A smooth (or analytic) *loop* on a manifold $M$ is a smooth (or analytic) map

$$M \times M \to M \; : (x, y) \mapsto x \circ y$$

which admits a unit $e$, i.e. $e$ is an element of $M$ such that $e \circ x = x \circ e = x$ for every $x$ in $M$, and such that equations $a \circ x = b$ and $y \circ a = b$ have solutions $x = a \backslash b$ and $b/a$ which depend smoothly (or analytically) on $(a, b)$. The simplest loops are Lie groups. A loop is a Lie group if and only if it is associative.

*a) From loops to 3-webs.*

We suppose that $(M, \circ)$ is a smooth loop. We associate to it a 3-web on $M \times M$ as follows: we denote by $(x, y)$ a generic point on $M \times M$; we take for foliation $\mathcal{F}_1$ the "horizontal" foliation given by $y =$ constant, for $\mathcal{F}_2$ we take the "vertical" foliation $x =$ constant, for $\mathcal{F}_3$ we take the foliation $x \circ y =$ constant.

*b) From 3-webs to loops.*

We consider a 3-web as above. We fix a point $m$ on $V$ and we denote by $M$ the leaf of $\mathcal{F}_1$ passing at $m$. We obtain a smooth loop on a neighborhood of $m$ in $M$ using the method illustrated in the figure 1.

In fact we obtain here only a "local" loop: we have three maps

$$(x, y) \mapsto x \circ y, \quad (x, y) \mapsto x/y, \quad (x, y) \mapsto y \backslash x$$

defined on a neighborhood of $m$ with $m \circ x = x \circ m = x$ and $(y/x) \circ x = y = x \circ (x \backslash y)$ every time the different members make sense. Note that, to get such a local smooth loop, we need only to have a smooth map $(x, y) \mapsto x \circ y$ with a unit; the implicit function theorem gives the two other maps in a neighborhood of this unit.

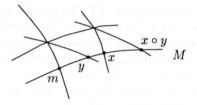

Figure 1.

## 3 Local normal form for loops.

**Theorem 1.** *Let* $(x, y) \mapsto x \circ y$ *be a smooth or analytic loop. Near the unit we have local coordinates* $(x_1, \ldots, x_n)$, *vanishing at the unit, such that*

$$x \circ y = x + y + \epsilon(x, y)$$

$$(x = (x_1, ..., x_n), y = (y_1, ..., y_n))$$

*where* $\epsilon$ *verifies*

$$\epsilon(x, 0) \equiv \epsilon(0, y) \equiv \epsilon(x, x) \equiv 0.$$

*Proof.* For every coordinate system $(x_1, \ldots, x_n)$ which vanishes at the unit, the function $f(x, y) := x \circ y$ verifies $f(x, 0) \equiv x$ and $f(0, y) \equiv y$ because 0 is the unit. We get then $f(x, y) = x + y + \nu(x, y)$ with $\nu(x, 0) \equiv \nu(0, y) \equiv 0$.

So we have $f(x, x) = 2x + 0(\|x\|)$. The classical linearization theorem of Sternberg [7] gives the commutative diagram

$$
\begin{array}{ccc}
\mathbb{R}^n, 0 & \xrightarrow{x \to f(x,x)} & \mathbb{R}^n, 0 \\
{\scriptstyle h}\downarrow & & \downarrow{\scriptstyle h} \\
\mathbb{R}^n, 0 & \xrightarrow{x \to 2x} & \mathbb{R}^n, 0
\end{array}
$$

where $h$ is a local diffeomorphism. So we obtain that $f'(x, y) = hf\left(h^{-1}(x), h^{-1}(y)\right)$ has the form $f'(x, y) = x + y + \epsilon'(x, y)$ with an $\epsilon'$ verifying $\epsilon'(x, 0) \equiv \epsilon'(0, y) \equiv \epsilon'(x, x) \equiv 0$. So we obtain the result $\qquad \square$

We consider now two smooth (or analytic) loops $(M, \circ)$ and $(M', \circ')$. An homomorphism from the first to the second is a map $h : M \to M'$ such that $h(x) \circ' h(y) = h(x \circ y)$. We see that $h$ maps the unit of $M$ to the unit of $M'$.

**Lemma 1.** *Let $\circ$ and $\circ'$ two (local) smooth or analytic loops in normal forms. This means that they are respectively defined on neighborhoods of the origin in $\mathbb{R}^n$ and $\mathbb{R}^p$ with*

$$x \circ y = x + y + \epsilon(x,y), \quad x \circ' y = x + y + \epsilon'(x,y)$$

*with*

$$\epsilon(x,0) \equiv \epsilon(0,y) \equiv \epsilon(x,x) \equiv 0, \quad \epsilon'(x,0) \equiv \epsilon'(0,y) \equiv \epsilon'(x,x) \equiv 0.$$

*Let $h$ be an homomorphism from the first loop to the second. If $h$ is differentiable at the origin then $h$ is linear.*

*Proof.* We have the relation

$$h(x) + h(y) + \epsilon'(h(x), h(y)) = h(x + y + \epsilon(x,y)).$$

When we impose $x = y$ this relation becomes $2h(x) = h(2x)$; and we obtain $h(x) = 2^n h\left(\frac{x}{2^n}\right)$ for every $x$ near enough $0$ and all $n$. We write $h(x) = L(x) + \|x\|\eta(x)$ with $L$ linear and $\eta(x) \xrightarrow[x \to o]{}$ ; so we have

$$L(x) + \|x\|\eta(x) = L(x) + \|x\|\eta\left(\frac{x}{2^n}\right)$$

for every $n$; then we obtain $\|x\|\eta(x) \equiv 0$ and $h(x) = L(x)$. This proves the result $\qquad\square$

**Corollary 1.** *Every homomorphism of smooth (or analytic) loops which is differentiable at the unit is smooth (or analytic) in a neighborhood of this unit.*

*Remarks.* The Campbell-Hausdorff formula $x \circ y = x + y + \frac{1}{2}[x,y] + \frac{1}{12}([[x,y],x] + \cdots) + \cdots$ is a normal form in the sense of our theorem 1 for every Lie group. The lemma 1 gives the interesting corollary: if we have local coordinates $x = (x_1, \ldots, x_n)$ near the unit of a Lie group $G$, vanishing at this unit, such that the law of $G$ is of the form $x \circ y = f(x,y)$ with $f(x,x) \equiv 0$, then $f(x,y)$ is the second member of the Campbell-Hausdorff formula. More rapidly: $f(x,x) \equiv 0$ characterizes the Campbell-Hausdorff formula.

## 4   Local normal form for 3-webs.

We return now to the 3-web $(\mathcal{F}_1, \mathcal{F}_2, \mathcal{F}_3)$. Fix a point $m$; near this point there are local coordinates $(x,y) = (x_1, \ldots, x_n, y_1, \ldots, y_n)$, vanishing at $m$, such that $\mathcal{F}_1$ and $\mathcal{F}_2$ are respectively $y =$ constant and $x =$ constant. Following strategy of section 2, we can suppose that $\mathcal{F}_3$ is given by $x \circ y =$ constant, for some local loop $x \circ y$. Then theorem 1 and lemma 1 have the corollary:

**Theorem 2.** *Near every point of a smooth or analytic 3-web $(\mathcal{F}_1, \mathcal{F}_2, \mathcal{F}_3)$ there are local coordinates $(x, y) = (x_1, \ldots, x_n, y_1, \ldots, y_n)$, vanishing at this point, such that the three foliations are respectively given by $y =$ constant, $x =$ constant and $x + y + \epsilon(x, y) =$ constant with*

$$\epsilon(x, 0) \equiv \epsilon(0, y) \equiv \epsilon(x, x) \equiv 0.$$

*Moreover these coordinates are unique up to a linear isomorphism of the form $(x, y) \mapsto (h(x), h(y))$.*

**Definition.** Let $W = (\mathcal{F}_1, \mathcal{F}_2, \mathcal{F}_3)$ and $W' = (\mathcal{G}_1, \mathcal{G}_2, \mathcal{G}_3)$ be two 3-webs respectively on the manifolds $V$ and $V'$. An *homomorphism from $W$ to $W'$* is a map $f : V \to V'$ which sends every leaf of $\mathcal{F}_i$ in a leaf of $\mathcal{G}_i$ for $i = 1, 2, 3$.

Corollary 1 gives following result.

**Theorem 3.** *An homomorphism of smooth (or analytic) 3-webs which is differentiable at a point is smooth (or analytic) in a neighborhood of this point.*

*Proof.* Call the homomorphism $f$ and $m$ the point under consideration. We choose normal forms for the two webs near $m$ and $f(m)$ like in theorem 2. The fact that $f$ preserves the first and second foliations gives $f(x, y) = (h(x), k(y))$. The fact that $f$ preserves the last foliation and the equations $\epsilon(x, 0) \equiv \epsilon(0, y) \equiv 0$ give $h = k$. Finally the fact that $f$ preserves the last foliation gives also that $h$ is an homomorphism of the loops $(x, y) \mapsto x + y + \epsilon(x, y)$ given by normal forms. Then we have only to apply corollary 1 to obtain the result $\qquad\qquad\square$

Theorems 1 and 2 first appeared in [4], in a more general form. At that time I was unaware of [1] where Akivis had given the "formal" analogue (the normal form for formal loops). Later, and apparently independently, Akivis and Shelekhov [2], have obtained the normal form for the analytic case.

Because of its unicity, the normal form contains local invariants for webs such that Akivis algebras, Chern curvatures and torsions (see, for example [3]).

When we work on 2-dimensional connected manifolds (real or complex), our results have the corollary that an homomorphism of 3-webs is completely determined by its behaviour on a pair of points (a linear map $(x, y) \mapsto (h(x), h(y))$ is determined by the image of a point different from the origin).

## 5  Rigidity of 3-webs

We consider a 3-web $W = (\mathcal{F}_1,\ \mathcal{F}_2,\ \mathcal{F}_3)$ on $V$. Fix a point $m$ on $V$. For every point $p$, near enough to $m$, we can construct a diffeomorphism $\phi_{m,p}$ from a neighborhood of $m$ to a neighborhood of $p$ as shown on the figure 2.

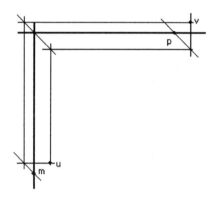

Figure 2.

In this figure verticals and horizontals represent respectively leaves of $\mathcal{F}_2$ and $\mathcal{F}_1$, transversal lines represent leaves of $\mathcal{F}_3$. The map $\phi_{m,p}$ sends $m$ to $p$ and $u$ to $v$. Now suppose that $f$ is an homomorphism from $W$ to another 3-web $W'$ which sends $m$ on $m'$ and $p$ on $p'$. By construction we have a relation $f \circ \phi_{m,p} = \phi_{m',p'} \circ f$ when we restrict conveniently $f$ and $\phi_{m,p}$. As a corollary we see that, if $f$ is continuous at $m$, it is continuous at $p$. This technic leads to the following lemma.

**Lemma 2.** *Let $f$ be an homomorphism between two 3-webs on connected manifolds. If $f$ is continuous at a point it is continuous everywhere.*

In fact the same method proves that, if $f$ is differentiable at a point, it is differentiable everywhere; but theorem 2 gives the better result that $f$ is smooth (or analytic). Now we are able to prove our rigidity theorem.

**Theorem 4.** *Every homomorphism between two 3-webs which is continuous is automatically smooth (or real analytic if the webs are analytic).*

*Proof.* We will prove first that $f$ is locally Lipschitz. To do this we fix a point $m$ in the first manifold. We choose coordinates in neighborhoods of $m$ and $f(m)$ which give to the two webs the normal form of theorem 2. Also we can suppose, as in the proof of theorem 3, that $f$ has the form $(x, y) \mapsto (h(x), h(y))$ where $h$ is a (continuous) homomorphism of the corresponding loops o and o'. Then we can write, for $x$ different from $y$ and $h(x)$ different from $h(y)$.

$$\frac{\|h(x) - h(y)\|}{\|x - y\|} = \frac{\|h(z)\|}{\|z\|} \frac{A}{B} \frac{\|h(x) - h(y)\|}{\|x - y\|} = \frac{\|h(z)\|}{\|z\|} \frac{A}{B}$$

with $z$ different from 0 and

$$A = \frac{\|h(x) - h(y)\|}{\|h(z)\|}, \quad B = \frac{\|x - y\|}{\|z\|}.$$

Now we take $z = y \backslash x$, which is equivalent to $y \circ z = x$. This point $z$ is uniquely defined at least if $x$ and $y$ are sufficiently near the origin. We can write then

$$B = \frac{\|y \circ z - y\|}{\|z\|} = \frac{\|z + \epsilon(y, z)\|}{\|z\|},$$

and this shows that we have, near the origin, $B > 1/2$. We have also

$$A = \frac{\|h(y \circ z) - h(y)\|}{\|h(z)\|} = \frac{\|h(y) \circ' h(z) - h(y)\|}{\|h(z)\|} = \frac{\|h(z) + \epsilon'(h(y), h(z))\|}{\|h(z)\|},$$

and the continuity of $h$ leads to $A < 2$, near the origin. Now as $h$ is an homomorphism we get (as in the proof of lemma 1) $h(2x) = 2h(x)$ which leads to

$$\frac{\|h(z)\|}{\|z\|} = \frac{\|h(z/2^n)\|}{\|z/2^n\|}.$$

So the first member is bounded on a disc $\|z\| \le r$ if and only if it is bounded on the annulus $r/2 \le \|z\| \le r$. But this is a consequence of the continuity of $h$.

These results show that $\|h(x) - h(y)\|/\|x - y\|$ is bounded on a neighborhood of the origin and prove that $h$ is locally Lipschitz.

Now recall the Rademacher's result [5] which says that a locally Lipschitz function is almost everywhere differentiable. Then our result follows from the theorem 3 $\qquad\square$

In this theorem we could replace the hypothesis "continuous" by "measurable" and keep the same conclusion. However the Axiom of Choice gives the existence of $\mathbb{Z}$-linear maps which are not measurable: they lead to homomorphisms of the 3-web $x$ =constant, $y$ =constant, $x + y$ =constant which are not measurable.

If we deal with holomorphic 3-webs the results of sections 2,3 and 4 are still true (replace analytic by holomorphic). However there is a little difficulty with theorem 4: it shows only that a continuous homomorphism between holomorphic 3-webs is real analytic. Using an interesting "holonomy" method, I. Nakai showed in [6] that, for a non linearizable holomorphic 3-web in $\mathbb{C}^2$, every continuous isomorphism is holomorphic or anti-holomorphic. We can prove this directly using our normal form and the fact that an isomorphism of these normal forms is $\mathbb{R}$-linear.

## References

1. M. A. AKIVIS, *Canonical expansions for Quasigroups*, Doklady Akad. Nauk. SSSR **188** (1969), 967–940.

2. M. A. AKIVIS & A. M. SHELEKHOV, *Canonical coordinates in a local analytic loop*, in Webs and Quasigroups (en russe), Kalinin Gos. Univ., Kalinin, (1986), 120–124, 130.

3. J. P. DUFOUR, *Introduction aux tissus, Seminaire Gaston Darboux*, Montpellier (1989/90).

4. J. P. DUFOUR & P. JEAN, *Rigidity of webs and families of hypersurfaces*, in Singularities and Dynamical Systems (Iraklion, 1983), North-Holland, Amsterdam, (1985), 271–283.

5. A. FRIEDMAN, *Differential Games*, Wiley Interscience, New-York, 1971.

6. I. NAKAI, *Topology of complex webs of codimension one and geometry of projective space curves*, Topology **26** (1987), no. 4, 475–504.

7. S. STERNBERG, *On the structure of local homeomorphisms of euclidean n-space II*, Am. J. of Math. **80** (1958), 623-631.

# RESONANT GEOMETRIC OPTICS AND WEBS *

JEAN-LUC JOLY

*MAB, Université Bordeaux I, 33405 Talence, France, joly@math.u-bordeaux.fr*

GUY MÉTIVIER

*IRMAR, Université Rennes I, 35042 Rennes, France,
Guy.Metivier@univ-rennes1.fr*

JEFFREY RAUCH

*Department of Mathematics, University of Michigan, Ann Arbor 48109 MI, USA,
rauch@math.lsa.umich.edu*

Geometric optics yields approximate solutions of Maxwell's equations of physic optics. The latter theory deals with oscillatory waves from Gamma radiation in radioactivity with wavelength $\approx 10^{-7}m$ to long radio waves ($\approx 10^3 m$) through X-rays and light waves. Geometrical optics is an approximation which ignores a part of the wave character of the electromagnetic light waves, and has propagation along rays. It furnishes a reasonable prediction as long as the wave length is suitably small. Thus geometrical optics is a high-frequency approximation.

Luneburg [12] was one of the first to present a clear link between geometric optics and the Maxwell equations. The mathematical validity of the method was established by Lax in a paper [10] which is a milestone in the modern theory of linear partial differential equations.

We first describe briefly the Lax-Luneburg method in the case of linear optics as background for the nonlinear case where webs play a role.

Consider the symmetric hyperbolic differential operator

$$L := \partial_t + \sum_{1 \le j \le d} A_j(t,x)\, \partial_{x_j}, \tag{1}$$

where $t \in \mathbb{R}$ denotes the time and $x_j$, $j = 1, \ldots, d$ the space coordinates. Symmetric hyperbolicity which means that the $A_j$ are smooth symmetric matrix valued functions. We look for solutions to

$$Lv^\varepsilon = g^\varepsilon \tag{2}$$

*RESEARCH PARTIALLY SUPPORTED BY THE U.S. NATIONAL SCIENCE FOUNDATION, U.S. OFFICE OF NAVAL RESEARCH, AND THE NSF-CNRS CO-OPERATION PROGRAM UNDER GRANTS NUMBER NSF-DMS-9203413 AND OD-G-N0014-92-J-1245 NSF-INT-9314095 RESPECTIVELY, AND THE CNRS THROUGH THE GROUPE DE RECHERCHE G1180 POAN.

of the form

$$u^\varepsilon(t,x) \sim \sum_{k \in \mathbb{N}} \varepsilon^k \, a_k(t,x) \, e^{i\,\varphi(t,x)/\varepsilon}, \tag{3}$$

that is oscillatory solutions with $\varphi$ a real valued phase function and frequency $\sim 1/\varepsilon$, $\varepsilon > 0$ being a small parameter. The source term is supposed to have the same form

$$g^\varepsilon(t,x) \sim \sum_{k \in \mathbb{N}} \varepsilon^k \, b_k(t,x) \, e^{i\,\varphi(t,x)/\varepsilon}.$$

The symbol of (1) is

$$A(t,x,\xi) := \sum_{1 \le j \le d} \xi_j \, A_j(t,x). \tag{4}$$

Introducing eigenvalues and eigenprojectors this symmetric matrix is written as

$$A(t,x,\xi) = \sum_\ell \lambda_\ell(t,x,\xi) \, \Pi_\ell(t,x,\xi). \tag{5}$$

With an assumption of constant multiplicity, the real eigenvalues $\lambda$ and the orthogonal spectral projectors $\Pi$ appearing in (5) are $C^\infty$ functions of $\xi \ne 0$.

The construction of the approximate solution is different according to whether $\varphi$ is *characteristic* for (1) or not, which means that the matrix

$$L(d\varphi) = \partial_t \varphi + A(t,x,\partial_x \varphi) \tag{6}$$

is singular or not. In nonlinear optics, where sums of phase functions appear naturally from products of exponentials, the property that a sum of characteristic functions is characteristic is closely related to the existence of abelian relations on a web.

Returning to our linear problem, we look for $a_0$ and $a_1$ such that

$$L(\partial) \left( (a_0 + \varepsilon a_1) \, e^{i\,\varphi/\varepsilon} \right) = b_0 \, e^{i\,\varphi/\varepsilon}, \tag{7}$$

which gives

$$\frac{1}{\varepsilon} L(id\varphi) \, a_0 + L(id\varphi) \, a_1 + L(\partial) \, a_0 = b_0, \qquad \varepsilon > 0.$$

Equating with 0 the coefficients of the successive power of $\varepsilon$ leads to

$$L(id\varphi) \, a_0 = 0, \qquad L(id\varphi) \, a_1 + L(\partial) \, a_0 = b_0 \tag{8}$$

If $L(id\varphi)$ is invertible at every point of an open set $\omega$ that is if $\varphi$ is noncharacteristic for (1) in $\omega$ then (8) reads

$$a_0 = 0, \qquad a_1 = L(d\varphi)^{-1} b_0.$$

and we say there is an elliptic inversion. Note that our approximate solution is of order $\varepsilon$ whereas the source term is of order 1 in $\varepsilon$.

Consider next the opposite extreme where $\det(L(id\varphi)) \equiv 0$ in $\omega$. If $d\varphi \neq 0$, this is equivalent to the fact that $\varphi$ satisfies in $\omega$ one and only one of the so-called eikonal equations

$$\partial_t \varphi + \lambda(t, x, \partial_x \varphi) = 0.$$

Here $\lambda$ denotes one of the eigenvalues in (5). Thus $a_0$ in (8) needs not be 0 but must satisfy the polarization condition

$$a_0 = \Pi(id\varphi) a_0, \tag{9}$$

where $\Pi$ is the projector associated to $\lambda$. Next set the coefficient of $\varepsilon^0$ in (8) equal to zero to obtain

$$L(id\varphi) a_1 + L(\partial) a_0 = b_0. \tag{10}$$

Since $L(id\varphi)$ is not invertible, (10) first implies $L(\partial) a_0 - b_0 \in \operatorname{Im} L(id\varphi)$ which is equivalent to the solvability condition

$$\Pi(id\varphi) L(\partial) a_0 = \Pi(id\varphi) b_0.$$

It follows that without trying to solve (10) completely we have obtained a second equation for $a_0$, which, taking (9) into account, reads

$$\Pi(id\varphi) L(\partial) \Pi(id\varphi) a_0 = \Pi(id\varphi) b_0. \tag{11}$$

This new equation is symmetric hyperbolic on the space of functions whose values satisfy $\Pi(id\varphi) a = a$. In fact one can prove that this operator is "scalar". Its principal symbol is equal to

$$\partial_t + \lambda'_\xi(t, x, d\varphi(t, x))(\partial_x)$$

and introduces the "group velocity" $\partial_\xi \lambda(d\varphi)$. Its characteristic variety is the tangent bundle of the characteristic variety of the operator $L(\partial)$ at $d\varphi$. It follows that the Cauchy problem for (11) is well posed in $C^\infty$ thus defining a regular $a_0 = \Pi(id\varphi) a_0$, given an arbitrary regular $a_{0|t=0} = \Pi(id\varphi) a_{0|t=0}$.

In summary, the resolution of

$$Lv^\varepsilon = b_0 \, e^{i\varphi/\varepsilon} \tag{12}$$

leads to the following pseudo-alternative.

1) If the $\varphi$ in the source term of (12) is not characteristic, then the inversion in (12) is elliptic and

$$v^\varepsilon \sim \varepsilon a_1 \, e^{i\varphi/\varepsilon}, \qquad a_1 = L(i\,d\varphi)^{-1}b_0 \qquad (13)$$

is $O(\varepsilon)$.

2) If $\varphi$ is characteristic then the solution to (12) is given by

$$
\begin{aligned}
&v^\varepsilon \sim a_0 \, e^{i\varphi/\varepsilon}, \quad \Pi(id\varphi)\,L(\partial)\,\Pi(id\varphi)a_0 = \Pi(id\varphi)\,b_0,\\
&a_{0|t=0} = \Pi(id\varphi)\,a_{0|t=0}
\end{aligned}
\qquad (14)
$$

thus has amplitude $O(1)$ and propagates with the group velocity.

The algorithm can be continued computing the infinite sequence of the $a_k$ in (3), each $a_k$ depending on the $a_j, j < k$, thus obtaining an approximate solution of the form (3) up to a $O(\varepsilon^{-\infty})$ error.

The analysis of the nonlinear problem

$$Lv^\varepsilon = f(v^\varepsilon), \qquad (15)$$

looking for approximate high-frequency oscillatory solutions with just one phase function $\varphi$ can be treated similarly. Since nonlinearities create harmonics, solutions are naturally sought in the form

$$u^\varepsilon = \sum_{k,\ell} \varepsilon^k a_{k,\ell}(t,x)\, e^{i\,\ell\varphi/\varepsilon} = \sum_k \varepsilon^k \, U_k(t,x,\varphi/\varepsilon), \qquad (16)$$

with $U_k \in C^\infty(\mathbb{R} \times \mathbb{R}^d \times \mathbb{R}/2\pi\mathbb{Z})$.

The relation between webs and optics occurs only when several phases are in play leading to the so-called three or more wave resonant interaction in physics. Approximate solutions are sought in the form

$$u^\varepsilon = \sum_k \varepsilon^k \, U_k(t,x,\varphi_1/\varepsilon,\ldots,\varphi_m/\varepsilon),$$

the functions $U_k$ being periodic in the $\theta_1,\ldots,\theta_m$ variables, and the phases $\varphi_1,\ldots,\varphi_m$ are $\mathbb{Q}$-linearly independent. The principal term $U_0(t,x,\varphi_1/\varepsilon,\ldots,\varphi_m/\varepsilon)$ is often written as a superposition of an often infinite number of waves of the form (16)

$$U_0(t,x,\varphi_1/\varepsilon,\ldots,\varphi_m/\varepsilon) = \sum_j U_j(t,x,\psi_j/\varepsilon)$$

where phases $\psi_j$ are linear combinations of the $\varphi_1,\ldots,\varphi_m$ and supposed to be *characteristic*.

Nonlinear interactions thus introduce source terms of the form

$$a(t,x)\, e^{i\,(\alpha_1\psi_1+\cdots+\alpha_k\psi_k)/\varepsilon}$$

which in turn lead to a $O(1)$ solution whenever

$$\psi = \alpha_1\psi_1 + \cdots + \alpha_k\psi_k \tag{17}$$

is characteristic. The equality (17) is called a resonance relation or phase matching. Resonant optics describes such phenomena. In mathematical terms, (17) is an abelian relation of the codimension 1 web defined on $\mathbb{R}^{1+d}$ by the functions $\psi$ and $\psi_k$. An aim of this paper is to describe some examples from resonant geometric optics which show how to use the theory of webs and also pose new questions for that theory.

## 1   Approximate solutions in space dimension 1

In one dimensional space, resonant geometric optics approximations are fairly simple and have a close direct connection with the geometry of plane webs.

Consider a general $3 \times 3$ semilinear strictly hyperbolic system,

$$X_k u_k^\varepsilon := \partial_t u_k^\varepsilon + \lambda_k(t,x)\,\partial_x u_k^\varepsilon = f_k(t,x,u_1^\varepsilon,u_2^\varepsilon,u_3^\varepsilon), \quad k=1,2,3. \tag{18}$$

The $\lambda_k(t,x)$ are smooth real-valued with $\lambda_k(t,x) \neq \lambda_{k'}(t,x)$ at every point of some open subset of $\mathbb{R}^{1+3}$. The $f_k$ are smooth functions of their arguments.

For simplicity, oscillatory initial data are given with just one phase

$$u_k^\varepsilon{}_{|t=0} = A_k(x,\psi(x)/\varepsilon), \quad x \in \omega, \tag{19}$$

where $\omega$ is an interval, $\psi \in C^\infty(\omega;\mathbb{R})$, $d\psi \neq 0$, a.e. $x \in \omega$, and $A_k \in C^0(\omega \times \mathbb{R}/2\pi\mathbb{Z})$.

Introduce the solutions to

$$X_k \psi_k = 0, \quad \psi_{k|t=0} = \psi, \quad k=1,2,3,$$

in an open set of determinacy, $\Omega$, of $\omega$. The solution to (18), (19) when $f_k = 0$ is the superposition of three oscillatory waves with $\psi_k$ as phases. In the nonlinear case, the 3 modes interact.

We prove that $u_k^\varepsilon(t,x) = \mathcal{U}_k(t,x,\psi_k(t,x)/\varepsilon) + o(1)$ with $\mathcal{U}_k$ the unique solutions to the profile equations of geometric optics:

$$X_k U_k(t,x,\theta) = (E_k F_k)(t,x,\theta), \tag{20}$$

$$U_k(0,x,\theta) = A_k(x,\theta) \tag{21}$$

Here

$$U_k \in C(\Omega \times \mathbb{R}/2\pi\mathbb{Z}),$$
$$F_k(t, x, \theta_1, \theta_2, \theta_3)) = f_k(t, x, U_1(t, x, \theta_1), U_2(t, x, \theta_2), U_3(t, x, \theta_3))$$

and $E_k$ is an integral operator which takes into account possible resonances between the phases $\psi_k$. Since the space dimension is 1, we have $\lambda(t, x, \xi) = \lambda(t, x)\, \xi$ so that the profiles $U_k$ propagate with the group velocity.

**Definition.** *A resonance on an open set $\Omega$ is a triple $(\varphi_1, \varphi_2, \varphi_3)$ of smooth functions satisfying $d\varphi_k \neq 0$ on $\Omega$ and the equations*

$$X_k \varphi_k = 0, \quad k = 1, 2, 3, \tag{22}$$

$$\sum_{1 \leq k \leq 3} d\varphi_k = 0. \tag{23}$$

Resonances do not depend on the $X_k$ but rather on the foliations –the 3-web– they define on $\Omega$. Resonances are the abelian relations of the web and the vector space of resonances (modulo constants) has dimension 0 or 1. There exists a resonance between oscillatory modes with phases $\psi_k$ if and only if the vector fields $X_k$ satisfy a strong condition and the $\psi_k$ are exactly the resonant phases. Constant vector fields and linear characteristic phases are good examples.

To understand the structure of the operators $E_k$, first observe that the nonlinearities $f_k$ create oscillations with all the phases $\sum \alpha_k \psi_k$ and that, for a fixed $1 \leq j \leq 3$, only combinations satisfying $X_j(\sum \alpha_k \psi_k) = 0$ will propagate as explained earlier. Since we assumed that the $\psi_k$ have the same trace $\psi$ when $t = 0$, the condition $X_j(\sum \alpha_k \psi_k) = 0$ implies there exists $\beta_j$ such that $\sum \alpha_k \psi_k = \beta_j \psi_j$. Thus, changing the $\alpha_k$ if necessary, for a resonance with phases $\psi_k$ to exist it is necessary that $\sum \alpha_k \psi_k = 0$.

Introduce the set of *resonance relations* defined as

$$R := \left\{ \alpha \in \mathbb{R}^3; \sum \alpha_k \psi_k = 0 \right\}. \tag{24}$$

When $R = \{0\}$ set

$$(E_1 F)(t, x, \theta_1) = \int_0^{2\pi} \int_0^{2\pi} F(\theta_1, \theta_2, \theta_3) \frac{d\theta_2}{2\pi} \frac{d\theta_3}{2\pi},$$

with similar definition for other indices. When $R$ is not trivial it is one dimensional. Choose $\alpha \neq 0$ in $R$ and set

$$(E_1 F)(t, x, \theta_1) = \lim_{h \to \infty} \int_0^h F\left(\theta_1, -\frac{\alpha_1 \theta_1 + \alpha_3 \theta_3}{\alpha_2}, \theta_3\right) \frac{d\theta_3}{h}.$$

Note that $E_1$ is just the average of an almost-periodic function and that it does not depend on the choice of $\alpha$ in $R \setminus \{0\}$.

Nonlinear interactions are governed by the operators $E$. In the nonresonant case oscillations have effect only on the mean value. In the resonant case they may produce new oscillations. The validity of geometric optics relies on the following result.

**Theorem.** *Suppose that for all $\alpha \in \mathbb{R}^3$ one has either*

$$\sum_k \alpha_k d\psi_k(t, x) \neq 0, \quad a.e. \ (t, x) \in \Omega,$$

*or*

$$\sum_k \alpha_k d\psi_k(t, x) = 0, \quad for \ all \ (t, x) \in \Omega.$$

*Then there exists $T > 0$ such that,*

*$\imath$) for all $0 < \varepsilon \leq 1$ the Cauchy problem (18) (19) has a unique solution $u^\varepsilon$ in $\Omega_T = \Omega \cap \{0 < t < T\}$,*

*$\imath\imath$) equations (20) (21) have a unique solution $U = (U_1, U_2, U_3) \in C(\Omega_T \times \mathbb{R}/2\pi\mathbb{Z})$,*

*$\imath\imath\imath$) $u_k^\varepsilon - U_k(\,.\,, \psi_k/\varepsilon) \to 0$ in $L^1(\Omega_T)$ as $\varepsilon \to 0$.*

*Remarks and Questions.* 1) There is a simple way to decide whether a 3-web is resonant or not. Consider the 3 vector fields

$$X_1 = \partial_t, \quad X_2 = \partial_x, \quad X_3 = \partial_t + c(t, x) \partial_x.$$

According to (22), (23) the associated web is resonant if and only if there exist $\varphi_1 = \varphi_1(x)$, $\varphi_2 = \varphi_2(t)$ such that

$$\varphi_2'(t) + c(t, x)\varphi_1'(x) = 0.$$

Hence $c(t, x) = c_1(x)c_2(t)$ or equivalently

$$\partial_t \partial_x \ln(|c(t, x)|) = 0. \tag{25}$$

Conversely, (25) implies the existence of a resonance. The condition (25) can be invariantly defined using the so-called curvature [2], which in our example is

$$\kappa = c^{-1}(t,x)\, \partial_t\, \partial_x\, \ln(|c(t,x)|).$$

2) A more detailed exposition can be found in [6], in particular concerning the averaging operators and the properties of resonances. Many questions remain unsolved. For example for webs with more than three foliations we lack simple criteria to decide whether there exist resonances.

## 2 Trilinear compensated compactness in one space dimension

We discuss now some weak continuity properties of cubic functions that rely deeply on the resonance properties of a 3 web. Extension of the result to functions of degree 4 is open.

Let $\Omega$ be an open subset of $\mathbb{R}^d$. Given 2 sequences of vector-valued functions such that $u_1^\varepsilon \rightharpoonup 0$, $u_2^\varepsilon \rightharpoonup 0$ in $L_{loc}^2(\Omega)$ and a bilinear function $F$, in general $F(u_1^\varepsilon, u_2^\varepsilon)$ does not converge to 0 in $\mathcal{D}'(\Omega)$. Suppose now that $P(x,D)u_1^\varepsilon$ is bounded in $L_{loc}^2(\Omega)$, for some elliptic $P$ of order 1, then $u_1^\varepsilon$ converges strongly to 0 in $L_{loc}^2(\Omega)$ and $F(u_1^\varepsilon, u_2^\varepsilon)$ converges to 0 in $\mathcal{D}'(\Omega)$. Now define

$$\mathcal{W}(\Omega,P) := \{u = (u_1,\ldots,u_N) \in L_{loc}^2(\Omega)\ ;\ Pu \in L_{loc}^2(\Omega)\}.$$

where $P(x,D)$ is a first order differential operator, not necessarily elliptic. Tartar [16] showed it is possible to associate to $P$ a family of quadratic functions which are weakly continuous on $\mathcal{W}(\Omega,P)$. These particular products, "compatible" with $P$, can be described by an algebraic "null condition". A famous example is given by the wedge products of differential forms as being the only compatible products with exterior differentiation (the div-curl Lemma of Tartar). The interested reader can consult [14,16] and [3] for more details.

The same question concerning cubic functions is far more complex. We restrict ourself to the simple case where $P$ is a diagonal operator.

Let $\Omega$ be an open subset of $\mathbb{R}^{1+1}$ and $X_j = \partial_t + \lambda_j\, \partial_x$, $j = 1\ldots,N$ $N$ vector fields on $\Omega$, such that $\lambda_j(t,x) \neq \lambda_k(t,x)$ for all $j \neq k$ and $(t,x) \in \Omega$. Set

$$\mathcal{W}(\Omega,X) := \{u = (u_1,\ldots,u_N) \in L_{loc}^2(\Omega)\ ;$$
$$Xu = (X_1 u_1,\ldots,X_N u_N) \in L_{loc}^2(\Omega)\}.$$

For $N = 3$ we ask whether the product $u_1 \cdots u_N$ on $\mathcal{W}(\Omega, X)$ is weakly continuous with values in $\mathcal{D}'(\Omega)$.

For $N = 2$ the answer is easy as a consequence of a simple version of Tartar's result: $u_1 u_2$ is well defined on $\mathcal{W}(\Omega, X)$ with values in $L^2_{loc}(\Omega)$ and it is weakly continuous. In the particular case where $X_j u_j = 0$, this follows readily from Fubini's theorem.

When $N = 3$ the existence of a resonance is clearly an obstruction to the weak continuity of $u_1 u_2 u_3$. Let $\varphi_j$ satisfy (22), (23) then for smooth and compactly supported $a_j$ the functions

$$u_j^\varepsilon := a_j \, e^{i\,\varphi_j/\varepsilon}$$

define a sequence $u^\varepsilon = (u_1^\varepsilon, u_2^\varepsilon, u_3^\varepsilon)$ which converges weakly to 0 in $\mathcal{W}(\Omega, X)$ whereas $u_1^\varepsilon u_2^\varepsilon u_3^\varepsilon = a_1 a_2 a_3$ can be chosen $\neq 0$.

For $N = 3$ the converse is true. The lack of resonance is sufficient to insure that $u_1 u_2 u_3$ is well defined and weakly continuous. This result can be improved by showing that if the resonant phases are not present in some suitably defined "asymptotic spectrum" of the sequence $u^\varepsilon$ then $u_1 u_2 u_3$ is still continuous (see [13]).

**Theorem** [7]. *Let $X_j$, $j = 1, 2, 3$ be 2 by 2 linearly independent vector fields on an open subset $\Omega$ of $\mathbb{R}^2$ such that the curvature $\kappa$ of the associated 3 web satisfies $\kappa(y) \neq 0$ for some $y \in \Omega$. Then there exists $\omega$ an open neighborhood of $y$ in $\Omega$ such that $(u_1, u_2, u_3) \to u_1 u_2 u_3$ is well defined and weakly continuous in $\mathcal{W}(\omega, X)$ with values in $L^2_{loc}(\omega)$.*

*Sketch of proof.*

1) We first need a tool to detect oscillations in a bounded sequence of $L^2(\Omega)$ : the semi-classical defect measure (Wigner measure) of P. Gérard [4], P.-L. Lions-T. Paul [11]. See also the introductory exposition of N. Burq [1].

Let $q$ be in the Schwartz space $\mathcal{S}(T^*(\Omega))$ and

$$q(x, D)u(x) = \frac{1}{2\pi^d} \int_{\mathbb{R}^d} e^{i\,x\cdot\xi} q(x, \xi)\hat{u}(\xi) \, d\xi.$$

Given $u^\varepsilon \rightharpoonup u$ in $L^2(\Omega)$ introduce

$$(q(x, \varepsilon D)u^\varepsilon, u^\varepsilon) \tag{26}$$

where outer parentheses denote the complex scalar product in $L^2(\Omega)$ so that (26) is a kind of a microlocal second order moment.

When $u^\varepsilon = a(x)e^{i\varphi(x)/\varepsilon}$ one has, up to a strong null sequence in $L^2(\Omega)$,

$$q(x, \varepsilon D)\big(a(x)e^{i\,\varphi(x)/\varepsilon}\big) \sim q(x, d\psi(x))a(x)e^{i\,\varphi(x)/\varepsilon},$$

so that (26) for such a sequence converges to

$$\int_{\mathbb{R}^d} q(x, d\psi(x)) \, |a(x)|^2 \, dx = \int_{T^*(\mathbb{R}^d)} q(x, \xi) \, d\mu(x, \xi),$$

with

$$\mu = \int \delta(\xi - d\psi(x)) \, |a(x)|^2 \, dx.$$

This example shows how phases and amplitudes of oscillations are taken into account by (26). In general one has the following result.

**Proposition** [4,11]. *Given a bounded sequence $u^\varepsilon$ in $L^2(\Omega)$, there exists a subsequence $\varepsilon'$ and a positive Borel measure $\mu$ on $T^*(\mathbb{R}^d)$ such that for all $q \in S(T^*(\mathbb{R}^d))$,*

$$\lim_{\varepsilon \to 0} (q(x, \varepsilon D)u^{\varepsilon'}, u^{\varepsilon'}) = \int_{T^*(\mathbb{R}^d)} q(x, \xi) \, d\mu(x, \xi).$$

One then says that $\mu$ is a semi-classical defect measure of the sequence $u^\varepsilon$, which we denote by $\mu \in M_{sc}(u^\varepsilon)$. Sequences with a unique defect measure are said to be pure. For pure sequences we write $\mu = M_{sc}(u^\varepsilon)$.

Change of the independent variables in $u$ induce canonical transformations of $\mu$. One also introduces cross products $(q(x, \varepsilon D)u^\varepsilon, v^\varepsilon)$ and the associated, not necessarily positive, measures defining the set $M_{sc}(u^\varepsilon, v^\varepsilon)$. For each measure $\mu \in M_{sc}(u^\varepsilon, v^\varepsilon)$ there exists $\mu_1 \in M_{sc}(u^\varepsilon)$ and $\mu_2 \in M_{sc}(v^\varepsilon)$ such that

$$\mu \ll \sqrt{\mu_1 \mu_2} \,.$$

That is, $\mu$ is absolutely continuous with respect to both $\mu_1$ and $\mu_2$. This implies the following continuity result.

**Lemma 1.** *Let $u^\varepsilon$ and $v^\varepsilon$ be pure bounded sequences in $L^2(\mathbb{R}^2)$, one of them being supported in a fixed compact set. If the corresponding defect measures $\mu$ and $\nu$ are mutually singular on $\{\xi \neq 0\}$. Then for all $\sigma \in C_0^0(\mathbb{R}^d)$ vanishing on a neighborhood of $\{\xi = 0\}$, one has*

$$(\sigma(\varepsilon D)u^\varepsilon, v^\varepsilon) \to 0 \quad \text{as} \quad \varepsilon \to 0. \tag{27}$$

*The conclusion holds for all $\sigma$ when the support of $\mu$ or $\nu$ does not meet $\{\xi = 0\}$.*

2) We now sketch the main steps of the proof.

First we may assume that $X_1 = \partial_t$, $X_2 = \partial_x$ and $X_3 = \partial_t + c(t,x)\partial_x$ so that the curvature of the 3 web is

$$\kappa(t,x) = c^{-1}(t,x)\, \partial_t\, \partial_x \ln(|c(t,x)|).$$

To prove that $\int_\omega \varphi u_1^\varepsilon u_2^\varepsilon u_3^\varepsilon \to \int_\omega \varphi u_1 u_2 u_3$, it is sufficient to prove that $\int_\omega u_1^\varepsilon u_2^\varepsilon u_3^\varepsilon \to \int_\omega u_1 u_2 u_3$, when one of the sequences is compactly supported. Introduce a dyadic partition of unity $1 = \Delta_0(\xi) + \sum_{k\geq 1}\Delta(2^{-k}\xi)$, where $\Delta$ has support in $\frac{1}{2} \leq |\xi| \leq 2$. Denote $\Delta_k = \Delta(2^{-k}D)$. From $\sum_{k\in\mathbb{N}}\Delta_k u_j^\varepsilon = u_j^\varepsilon$, $j = 1,2,3$ one gets

$$\int_\omega u_1^\varepsilon u_2^\varepsilon u_3^\varepsilon = \sum_{i,j,k} \int_\omega \Delta_i\, u_1^\varepsilon\, \Delta_j\, u_2^\varepsilon\, \Delta_k\, u_3^\varepsilon.$$

In the right hand side of this equality every term $(i,j,k)$ in the sum such that $\sigma(i) * \sigma(j) < \sigma(j) + 3 * \sigma(k)$ for some permutation $\sigma$ is equal to 0 because 0 does not belong to the spectrum of $\Delta_i\, u_1^\varepsilon\, \Delta_j\, u_2^\varepsilon\, \Delta_k\, u_3^\varepsilon$. It is therefore sufficient to study sums of the form

$$I_{0,\alpha_2,\alpha_3} = \sum_{\substack{i\geq 0 \\ i+\alpha_2\geq 0 \\ i+\alpha_3\geq 0}} \int_\omega \Delta_i\, u_1^\varepsilon\, \Delta_{i+\alpha_2}\, u_2^\varepsilon\, S_{i+\alpha_3}\, u_3^\varepsilon, \quad |\alpha_2| \leq 2,\; |\alpha_3| \leq 2 \qquad (28)$$

where $S_k = \sum_{\ell\leq k}\Delta_\ell$. Each term in the series (28) is weakly continuous on $L^2(\Omega)^3$. The theorem will therefore be proved if we get a $L^2$ estimate on, say $I_{0,0,0}$, which is uniform in $\varepsilon$. We make this more precise.

**Proposition.** Let $X_j$ satisfy the assumption of the theorem. Suppose that $\sigma_j \in C_0^\infty(\mathbb{R}^2)$ and that one of them at least vanishes near 0. Then, for every compact $K \in \mathbb{R}^2$ there exists a function $\delta$ decreasing to 0 at infinity such that for all $v_j \in \mathcal{W}(\omega, X_j)$ with support in $K$

$$\left|\int_\omega \sigma_1(2^{-i}D)v_1\, \sigma_2(2^{-i}D)v_2\, \sigma_3(2^{-i}D)v_3\right| \leq \delta(i) \prod_{j=1,2,3} \|v_j\|_{W(X_j,\omega)}. \qquad (29)$$

The proof of the theorem follows from the above proposition setting $v_1 = \Delta_i u_1$, $v_2 = \Delta_i u_2^\varepsilon$ and $v_3^\varepsilon = S_i u_3^\varepsilon$, choosing $\sigma_j$ such that $\sigma_j(2^{-i}D)v_j^\varepsilon = v_j$ with $\sigma_1 = 0$ near 0 and noticing that

$$\|\Delta_i u\|_{W(X,\omega)} \leq C\, a_i(u)\, \|u\|_{W(X,\omega)}, \quad i \in \mathbb{N},\; \|a(u)\|_{\ell^2} \leq 1.$$

3) *Proof of the proposition.* Suppose that estimate (29) is false. There is a $\delta > 0$ and sequences $v_j^\varepsilon \in \mathcal{W}(\omega, X_j)$ with support in $K$ such that

$$|\int_\omega \sigma_1(\varepsilon D)v_1^\varepsilon \, \sigma_2(\varepsilon D)v_2^\varepsilon \, \sigma_3(\varepsilon D)v_3^\varepsilon| \geq \delta, \quad \text{and} \quad \|v_j^\varepsilon\|_{\mathcal{W}(\omega, X_j)} = 1.$$

It remains to show that this is in contradiction with Lemma 1. This follows from 3 Lemmas.

**Lemma 2.** *Suppose $X_1 = \partial_t$, $X_2 = \partial_x$ and $v_j^\varepsilon$ are bounded sequences in $\mathcal{W}(X_j, \omega)$, $j = 1, 2$. Then all measures in $M_{sc}(v_1^\varepsilon v_2^\varepsilon)$ are absolutely continuous with respect to the product measure $\nu_1(dx, d\xi) \otimes \nu_2(dt, d\tau)$.*

This result is easily checked in the special case $v_1^\varepsilon(t, x) = \psi_1(t)\psi_1^\varepsilon(x)$ and $v_2^\varepsilon(t, x) = \psi_2(x)\psi_2^\varepsilon(t)$.

**Lemma 3.** *Suppose $v_3^\varepsilon$ is bounded in $\mathcal{W}(X_3, \omega)$. Then every measure $\mu$ in $M_{sc}(v_3^\varepsilon)$ is carried by the characteristic variety of $X_3$ defined by*

$$C_3 := \{(t, \tau, x, \xi); \tau + c(t, x)\,\xi = 0\}$$

*Moreover for all $(x, \xi)$ and all $(t, \tau)$,*

$$\mu(C_3(x, \xi)) = \mu(C_3(t, \tau)) = 0$$

*where $C_3(x, \xi)$ denotes the section of $C_3$ over $(x, \xi)$ and similarly for $C_3(t, \tau)$.*

Lemma 2 and 3 and the nowhere-vanishing property of the curvature are used in the last step.

**Lemma 4.** *Assume that for all $(t, x) \in \omega$,*

$$\partial_t \partial_x \ln(|c(t, x)|) \neq 0.$$  $$(30)$$

*Then with the assumptions of Lemma 2 and 3, one has*

$$M_{sc}(v_1^\varepsilon v_2^\varepsilon) \perp M_{sc}(v_3^\varepsilon).$$

*Proof.* One shows that $\nu(C_3) = 0$ whenever $\nu \in M_{sc}(v_1^\varepsilon v_2^\varepsilon)$. As a consequence of Lemma 1, it is sufficient to show that $\int_{C_3} \nu_1(dx, d\xi) \otimes \nu_2(dt, d\tau) = \int \nu_1(C_3(t, \tau))\nu_2(dt, d\tau) = 0$. One may suppose that $\nu_1$ and $\nu_2$ have no atoms. In the integral it is sufficient to consider the set of $(t, \tau)$ such that $\nu_1(C_3(t, \tau)) > 0$. If one proves that for all $(t, \tau) \neq (t', \tau')$, one has

$$\nu_1\big(C_3(t, \tau) \cup C_3(t', \tau')\big) = \nu_1(C_3(t, \tau)) + \nu_1(C_3(t', \tau')),$$  $$(31)$$

it follows that there exists at most a countable set of $(t, \tau)$ such that $\nu_1(C_3(t, \tau)) > 0$. Hence $\int \nu_1(C_3(t, \tau))\nu_2(dt, d\tau) = 0$.

It remains to prove (31). This follows from the curvature condition (30) which implies that $C_3(t, \tau)$ and $C_3(t', \tau')$ cross transversally whenever $(t, \tau) \neq (t', \tau')$. Indeed if $(x, \xi) \in C_3(t, \tau) \cap C_3(t', \tau')$, the slopes of the 2 curves $C_3(t, \tau)$ and $C_3(t', \tau')$ in the plane $(x, \xi)$ are equal to $-\frac{\tau}{c^2(t, x)}\partial_x c(t, x)$ and $-\frac{\tau'}{c^2(t', x)}\partial_x c(t', x)$ respectively. Since $\frac{\tau}{c(t, x)} = \frac{\tau'}{c(t', x)}$, these slopes are equal only if $t = t'$ as follows from (30). This implies $\tau = \tau'$ which is impossible. Since $\nu_1$ has no atoms, (31) is proved.

## 3 Resonances in space of dimension greater than 1

1) **Some definitions.** Since $d > 1$, we cannot assume, as in the last sections, that the operator

$$L = \partial_t + \sum_{1 \leq j \leq d} A_j(t, x)\,\partial_j$$

has a diagonal form. Introduce

$$p_\ell(\tau, \xi) = \tau + \lambda_\ell(t, x, \xi), \qquad p(\tau, \xi) = \prod_\ell p_\ell(\tau, \xi), \qquad (32)$$

where the eigenvalues $\lambda_\ell$ of $A(t, x, \xi)$, are assumed to be distinct. Since $p(\tau, \xi) = |p(-\tau, -\xi)|$, the zeros of $p$ are symmetric with respect to $\tau = 0, \xi = 0$.

**Definition.** *For $p$ as in (32) a p-resonance on the open set $\Omega$ in $\mathbb{R}^{1+d}$ is an n-uple $\psi_1(t, x), \ldots, \psi_n(t, x))$ of smooth real functions on $\Omega$ such that*

$$\forall (t, x) \in \Omega, \; j = 1 \ldots n, \quad d\psi_j(t, x) \neq 0, \quad \text{and}$$

$$p(t, x, d\psi_k(t, x)) = 0, \quad 1 \leq k \leq n \qquad (33)$$

$$\sum_k d\psi_k = 0. \qquad (34)$$

The functions $\psi_k$ define an $n$-web of codimension 1 on $\Omega \subset \mathbb{R}^{1+d}$ and (34) is an abelian equation on the web. Property (33) requires that the conormal of each leaf annihilates $p$.

Conversely, suppose $\psi_k$, $k = 1 \ldots n$ are given such that $p(d\psi_k) = 0$. An abelian equation on the web defined by the $\psi_k$ is a relation

$$\sum_k f_k'(\psi_k)\,d\psi_k = 0$$

for non vanishing $f'_k$. Since $d(f_k(\psi_k)) \neq 0$ on $\Omega$ one immediately checks that the $\varphi_k := f_k(\psi_k)$ satisfy (33) and (34).

Thus (33) can be viewed as a null-$n$-web referring to the fact that each foliation satisfies (33). It also suggests a slightly more general setting in term of null lagrangians. If $S$ is a lagrangian such that $p_{|S} = 0$, then, outside caustics, $S$ locally defines a web of codimension 1 on $\Omega$.

In multi-d geometric optics we used the following definition.

**Definition.** *A linear space $\Phi(\Omega)$ of $C^\infty$ real-valued functions on $\Omega$ is said to be p-coherent if for any $\varphi \in \Phi(\Omega)$ one has either*

$$p(t, x, d\varphi(t, x)) \neq 0, \quad \forall (t, x) \in \Omega$$

*or*

$$d\varphi(t, x) \neq 0, \quad p(t, x, d\varphi(t, x)) = 0, \quad \forall (t, x) \in \Omega$$

Since any $\varphi \in \Phi(\Omega) \setminus \{0\}$ has nowhere vanishing differential, $\dim \Phi \leq d + 1$. As a typical example take linear functions and $p$ independent of $t$ and $x$.

2) *Remarks.*

a) When $d \geq 2$ there exists a resonance with only $d$ functions. This situation cannot occur when $d = 1$. We give an example due to Hunter, Majda and Rosales. Let $p = \tau(\tau^2 - |\xi|^2)$, $\xi \in \mathbb{R}^d$, $d \geq 2$. $p$ is the characteristic polynomial of the linearized Euler equations of gas dynamics around an equilibrium state with null speed, constant density and sound speed 1. The functions $\psi_1(t, x) = t - \alpha_1(x)$, $\psi_2(t, x) = \alpha_2(x) - t$ and $\psi_3(t, x) = \alpha_1(x) - \alpha_2(x)$ annihilate $p$ when $|\alpha'_1| = |\alpha'_2| = 1$ and they obviously satisfy

$$\sum_{1 \leq j \leq 3} \psi_j = 0.$$

We next discuss whether there exists another $p$-resonance with 3 functions.

Consider $\psi_3$ an arbitrary solution to $\partial_t \psi_3 = 0$ and look for $\psi_1$ and $\psi_2$ satisfying $\partial_t \varphi - |\partial_x \varphi| = 0$ such that

$$\psi_3 = \psi_1 - \psi_2. \tag{35}$$

Introduce

$$\psi = \psi_1 + \psi_2. \tag{36}$$

Since $\partial_x \psi_3 \cdot \partial_x \psi = |\partial_x \psi_1|^2 - |\partial_x \psi_2|^2 = |\partial_t \psi_1|^2 - |\partial_t \psi_2|^2 = \partial_t \psi_3 \partial_t \psi$, it follows that

$$\partial_x \psi_3 \cdot \partial_x \psi = 0 \tag{37}$$

Since $\partial_t \psi = |\partial_x \psi_1| + |\partial_x \psi_2| = \frac{1}{2}|\partial_x \psi_3 + \partial_x \psi| + \frac{1}{2}|\partial_x \psi_3 - \partial_x \psi|$, taking (37) into account, one gets $|\partial_x \psi_3 \pm \partial_x \psi| = \sqrt{|\partial_x \psi_3|^2 + |\partial_x \psi|^2}$ and thus

$$\partial_t \psi^2 = |\partial_x \psi_3|^2 + |\partial_x \psi|^2. \tag{38}$$

Conversely, if $\psi$ satisfies (37) and (38) the functions $\psi_1$ and $\psi_2$ defined by (35) et (36) are solutions to $\partial_t \varphi - |\partial_x \varphi| = 0$. We are thus led to look for $\psi$ satisfying (37) et (38), $\psi_3$ being considered as a parameter. Set

$$r = \psi_3'(x) \cdot \xi \tag{39}$$

$$s = \tau^2 - |\psi_3'(x)|^2 - |\xi|^2. \tag{40}$$

One checks

$$\{r, s\} = \psi_3''(x)(\psi_3'(x), \psi_3'(x)) - \psi_3''(x)(\xi, \xi). \tag{41}$$

Suppose now $d = 2$. We consider 2 cases.

$i$) The function $\psi_3$ satisfies $\psi_3''(x)(\psi_3'(x), \psi_3'(x)) \neq 0$ and $\psi_3''(x)(\xi, \xi) \neq 0$ for all $\xi \perp \psi_3'(x)$ and all $x \in \omega$. This is the generic case. The differential of a common solution $\psi$ to both equations (37) and (38) annihilates $r$ et $s$ and thus $\{r, s\}$. Since equations $r = 0$ et $\{r, s\} = 0$ uniquely determine $\partial_x \psi$ independent of $t$ one gets $\partial_t \partial_x \psi = 0$. Equation $s = 0$ then determines $\partial_t \psi$ independent of $t$ and also of $x$. Thus $\psi$ is of the form $\psi = 2\alpha t + \varphi(x)$ and it follows that $\psi_j = \alpha t - \varphi_j(x)$, $j = 1, 2$. These are the solutions of [5] discussed above.

$ii$) On the contrary suppose now that on $\omega$, $\psi_3$ satisfies

$$\psi_3''(x)(\psi_3'(x), \psi_3'(x)) \equiv 0. \tag{42}$$

In order that $\psi$ satisfies (37) and (38) it is sufficient that $\{r, s\} = 0$ on $r = 0$, or in other words that $\psi_3$ satisfies on $\omega$ the following condition

$$\psi_3''(x)(\xi, \xi) = 0 \quad \text{if} \quad \psi_3'(x)(\xi) = 0. \tag{43}$$

Solutions to (42) (43) are well known: they are either affine functions or the "angle" function $\theta$ whose graph is an helicoid. In contrast to the case $i$) $\partial_x \psi$ is no more uniquely determined by (42) and may depend on $t$. The functions $\psi_1$, $\psi_2$ are deduced from $\psi_3$ et $\psi$. Let us show that if $\psi_3$ is the "angle" function we recover new solutions different from those obtained in $i$). First observe that since $r = 0$, $\psi$ is radial in $x$, that is $\psi(t, x) = \psi(t, \rho)$. The second equation then implies that, on the determination $\tau = \sqrt{|\xi|^2 + \frac{1}{\rho^2}}$,

$$\partial_t \psi = \sqrt{|\partial_\rho \psi|^2 + \frac{1}{\rho^2}}.$$

Thus there exist resonances which are not affine functions of $t$.

b) Does focusing create resonance? Here is an example of a null lagrangian for $\tau^2 - |\xi|^2$, $\tau \in \mathbb{R}$, $\xi \in \mathbb{R}^2$. For more details see [9] and also [15] on this issue.

A first part of $S$ annihilates $\tau - |\xi|$ and when $t = 0$ is the graph of the differential of $\psi(y_1, y_2) = y_2 + y_1^2/2$. This phase function focuses at $t > 0$. $S$ is globally parameterized

$$
\begin{cases}
t = s, & \tau = (1 + y_1^2)^{1/2}, \\
x_1 = y_1 - s\, y_1/(1 + y_1^2)^{1/2}, & \xi_1 = y_1, \\
x_2 = y_2 - s/(1 + y_1^2)^{1/2}, & \xi_2 = 1.
\end{cases}
$$

The caustic set $\mathcal{C}$ is the projection of the subset $\mathcal{C}_S$ of $S$ whose equation in $(s, y)$ coordinates is

$$
s = (1 + y_1^2)^{3/2}.
$$

Thus

$$
\mathcal{C} = \{(t, x) \in \mathbb{R}^{1+2} \; ; \; t^{2/3} = x_1^{2/3} + 1, \; t \geq 1\}.
$$

Denote by $q$ the restriction of the canonical projection $T^*(\mathbb{R}^{1+2}) \to \mathbb{R}^{1+2}$ to $S$. The set $q^{-1}(\mathcal{C})$ is the union of $\mathcal{C}_S$ and a regular curve $\mathcal{S}_S$ given by

$$
s = \sigma(y_1)
$$

with $\sigma(y_1) \leq (1 + y_1^2)^{3/2}$ and equality only if $y_1 = 0$, $s = 1$. Points in $\mathcal{C}_S$ are singular for $q$ whereas $q$ is regular at all points of $\mathcal{S}_S \setminus \mathcal{C}_S$. Introduce

$$
G_0 := \{(t, y) : 0 \leq t < \sigma(y_1)\} \tag{44}
$$

and

$$
G_1 := \{(t, y) \; ; \; s(y_1) < t < (1 + y_1^2)^{3/2}, \; y_1 > 0\}
$$
$$
G_2 := \{(t, y) \; ; \; s(y_1) < t < (1 + y_1^2)^{3/2}, \; y_1 < 0\}
$$

and

$$
G_\infty := \{(t, y) ; (1 + y_1^2)^{3/2} < t\}.
$$

$q$ defines the diffeomorphism

$$
q : G_0 \to \Omega_0 := \{(t, x) ; 0 \leq t < (x_1^{2/3} + 1)^{3/2}\}
$$

and the diffeomorphisms

$$q : G_k \to \Omega_\infty := \{(t,x) : t > (x_1^{2/3} + 1)^{3/2}\}, \quad k = 1, 2, \infty.$$

Before the caustic, that is on $\Omega_0$, $q^{-1}(t,x)$ is one point, $(t, g_0(t,x))$, and $S$ defines one phase function

$$\varphi(t,x) = \psi(t, g_0(t,x)).$$

After the caustic, that is on $\Omega_\infty$, $q^{-1}(t,x)$ is 3-fold defining $(t, g_k(t,x)) \in G_k$, $k = 1, 2, \infty$, and 3 phases

$$\varphi_k(t,x) = \psi(t, g_k(t,x)), \quad k = 1, 2, \infty.$$

Write

$$(t, g_k(t,x)) = (t, y_1, y_2) := (t,\ h_{k,1}(t,x_1),\ x_2 + h_{k,2}(t,x_1)),$$

and observe that the phases are

$$\varphi_k(t,x) = x_2 + h_{k,2} + (h_{k,1})^2/2.$$

Their differentials are given by

$$d\varphi_k(t,x) = (1 + h_{k,1}^2)^{1/2} dt + h_{k,1} dx_1 + dx_2.$$

Note these differentials are 2 by 2 independent since the values of the $h_{k,1}$ are different. Since the $g_k$ can be extended continuously to $\overline{\Omega}_\infty$, the phases $\varphi_k$, $k = 0, 1, 2, \infty$ have $C^1(\overline{\Omega}_\infty)$ extensions. Introduce

$$C_1 := C \cap \{x_1 > 0\}, \quad \text{and,} \quad C_2 := C \cap \{x_1 < 0\}.$$

For $k = 1, 2$, $\varphi_k$ is the analytic extension of $\varphi_0$ through $C_k$. In particular, $\varphi_k = \varphi_0$, $d\varphi_k = d\varphi_0$ on $C_k$. Moreover $\varphi_1 = \varphi_\infty$, $d\varphi_1 = d\varphi_\infty$ on $C_2$, and, $\varphi_2 = \varphi_\infty$, $d\varphi_2 = d\varphi_\infty$ on $C_1$. At $t = t_0$ the graphs of $\varphi_\infty$, $\varphi_1$ et $\varphi_2$ form a swallow tail with cusps above $C_k \cap \{t = t_0\}$, $k = 1, 2$. The concave phase $\varphi_\infty$ gives the bottom of the tail, whereas $\varphi_1$ et $\varphi_2$ are convex boundaries extended by $\varphi_0$ on the right and the left side of the caustic. The line in $S$ with equation $t = t_0$, $y_2 = 0$ parameterized by $y_1$ is mapped on a curve parameterized by $x_1$ going from $+\infty$ to $C_2$ where $x_1$ starts decreasing, going from $C_2$ to $C_1$ where $x_1$ increases again, going from $C_1$ to $-\infty$.

**Proposition.** Let $\alpha_k$, $k = 1, 2, 3$ not all 0. If

$$\varphi = \alpha_1 \varphi_1 + \alpha_2 \varphi_2 + \alpha_\infty \varphi_\infty$$

then

$$\text{meas}\left(\{(\partial_t\varphi)^2 - |\partial_x\varphi|^2 = 0\} \cap \Omega_\infty\right) = 0.$$

*Proof.* Assume that $\omega := \{(\partial_t\varphi)^2 - |\partial_x\varphi|^2 = 0\} \cap \Omega_\infty$ has positive measure. Since the $g_k$ are analytic in $\Omega_\infty$, so is $\varphi$ thus $(\partial_t\varphi)^2 - |\partial_x\varphi|^2 = 0$ in $\Omega_\infty$. One also checks that $\varphi \in C^1(\overline{\Omega}_\infty)$.

On $\mathcal{C}_2$, $d\varphi = (\alpha_1 + \alpha_\infty)d\varphi_1 + \alpha_2 d\varphi_2$. Since $d\varphi$, $d\varphi_1$ and $d\varphi_2$ belong to $\tau^2 - |\xi|^2 = 0$ which is cut by an homogeneous plane along at most 2 lines, it follows that either $\alpha_1 + \alpha_\infty = 0$ or $\alpha_2 = 0$. A similar argument on $\mathcal{C}_1$ implies that either $\alpha_2 + \alpha_\infty = 0$ or $\alpha_1 = 0$. If $\alpha_1 = 0$ or $\alpha_2 = 0$ all $\alpha_k$ vanish in $\Omega_\infty$ since $d\varphi_k$ are independent 2 by 2. We are left to consider the case when $\alpha_1 + \alpha_\infty = \alpha_2 + \alpha_\infty = 0$, that is $\varphi = \varphi_1 + \varphi_2 - \varphi_\infty$. Compute $d\varphi(t,0,0)$ for $t > 1$. Since $d\varphi_\infty(t,0,0) = (1,0,1)$, $d\varphi_1(t,0,0) = (t,\sqrt{t-1},1)$ and $d\varphi_2(t,0,0) = (t,-\sqrt{t-1},1)$, one gets $d\varphi(t,0,0) = (2t-1,0,1)$ which is not a zero of $\tau^2 - |\xi|^2 = 0$ in $\Omega_\infty$ where $t > 1$. The contradiction completes the proof.

Concerning the same question, Gilles Robert in [15] performs a more general analysis. He shows that any fairly general algebraic Lagrangian variety of degree 4, annihilating $\tau^2 - |\xi|^2$, has the property that the 4-web it defines on the base has no abelian relation.

## References

1. N. BURQ, *Mesures semi-classiques et mesures de défaut*, Séminaire Bourbaki, 1996/1997, exposé no. 826, Astérisque, no. 245, (1997), 167–195.
2. W. BLASCHKE & G. BOL, *Geometrie der Gewebe*, Springer-Verlag, Berlin, 1938.
3. P. GÉRARD, *Microlocal Defect Measure*, Communications in Partial Differential Equations **16** (1991), 1761–1794.
4. P. GÉRARD, *Mesures semi-classiques et ondes de Bloch*, Séminaire Équations aux dérivées partielles, École Polytechnique, exposé no. 16, 1990/1991.
5. J. HUNTER, A. MAJDA & R. ROSALES, *Resonantly interacting weakly nonlinear hyperbolic waves II: several space variables*, Stud. Appl. Math. **75** (1986), 187–226.
6. J.-L. JOLY, G. MÉTIVIER & J. RAUCH, *Resonant one dimensional nonlinear geometric optics*, J. Funct. Anal. **114** (1993), no. 1, 106–231.
7. J.-L. JOLY, G. MÉTIVIER & J. RAUCH, *Trilinear compensated compactness*, Annals of Math. **142** (1995), 121–169.

8. J.-L. JOLY, G. MÉTIVIER & J. RAUCH, *Coherent and focusing multidimensional nonlinear geometric optics*, Annales de L'École Normale Supérieure de Paris **28** (1995), 51-113.

9. J.-L. JOLY, G. MÉTIVIER & J. RAUCH, *Nonlinear oscillations beyond caustics*, Comm. on Pure and Appl. Math., Vol. XLIX (1996), 443-527.

10. P. LAX, *Asymptotic solutions of oscillatory initial value problems*, Duke Math. J. **24** (1957), 627–645.

11. P.-L. LIONS & T. PAUL, *Sur les mesures de Wigner*, Revista Matemática Iberoamericana **9** (1993), no. 3, 553-618.

12. R. K. LUNEBURG, *Mathematical theory of optics*, Univ. of California Press, Berkeley, 1964.

13. G. MÉTIVIER & S. SCHOCHET, *Interactions trilinéaires résonnantes*, Séminaire Équations aux dérivées partielles, École Polytechnique, exposé no. 6, 1995/1996.

14. F. MURAT, *Compacité par compensation*, Ann. Scuola Norm. Sup. Pisa **5** (1978), no. 3, 489–507.

15. G. ROBERT, *A three-dimensional Lagrangian 4-web with no abelian relation*, this volume

16. L. TARTAR, *Compensated compactness and applications to partial differential equations*, Research Notes in Mathematics, vol. 4, New-York, Pitmann Press, 1979.

# INTRODUCTION TO $G$-STRUCTURES VIA THREE EXAMPLES

J.M. LANDSBERG

*Laboratoire de Mathématiques, Université Paul Sabatier, UFR-MIG, 31062*
*Toulouse Cedex 4, FRANCE*
*jml@picard.ups-tlse.fr*

**Introduction.** This is an expository article based on my lecture at the *Journées sur les tissus* at the Université Paul Sabatier, Toulouse, December 2-4, 1996. Its purpose is to introduce the reader to methods developed by E. Cartan for the study of geometric problems. The guiding principles are: 1. Do not make choices (e.g. coordinates) that have no geometric meaning (work on a bigger space instead). 2. When you don't know what to do next, differentiate.

## 1  First example: 3-webs in $\mathbb{R}^2$

### 1.1  First formulation of the question

Let $\mathcal{L} = \{L_1, L_2, L_3\}$ be a collection of three pairwise transverse foliations of $\mathbb{R}^2$ (or an open subset of $\mathbb{R}^2$, since we will work locally). Such a structure is called a *web*.

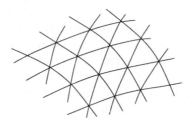

Figure 1.

Let $\tilde{\mathcal{L}} = \{\tilde{L}_1, \tilde{L}_2, \tilde{L}_3\}$ be another web.

**Question.** *When does there exist a (local) diffeomorphism* $\phi : \mathbb{R}^2 \to \mathbb{R}^2$ *such that* $\phi^*(\tilde{L}_j) = L_j$?

If there exists such a (local) $\phi$, we will say the webs $\mathcal{L}, \tilde{\mathcal{L}}$ are (locally) equivalent.

In particular, let $\mathcal{L}^0$ be the web $L_1^0 = \{y = const\}, L_2^0 = \{x = const\}, L_1^0 = \{y - x = const\}$. When is a web locally equivalent to $\mathcal{L}^0$ (the "flat" case)?

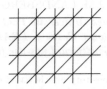

Figure 2.

## 1.2 Second formulation of the question

Let $y' = F(x, y)$ be an ordinary differential equation (henceforth called an *ODE*) in the plane. Let $y' = \tilde{F}(x, y)$ be another. When does there exist a change of coordinates $\psi : \mathbb{R}^2 \to \mathbb{R}^2$ of the form $\psi(x, y) = (\alpha(x), \beta(y))$ such that $\psi^* \tilde{F} = F$?

In particular, given $F$, is it equivalent to $y' = 1$ via a change of coordinates of the form $\psi$?

The two questions are the same: Any two transverse foliations can be used to give local coordinates $x, y$ and the space of integral curves of an ODE in coordinates provides the third foliation. The diffeomorphisms of $\mathbb{R}^2$ that preserve the two coordinate foliations are exactly those of the form of $\psi$.

In order to study the local equivalence of webs, we would like to associate a *frame* to a web, i.e., for all $x \in \mathbb{R}^2$ we would like to have a basis of $T_x^* \mathbb{R}^2$ (or equivalently, a basis of $T_x \mathbb{R}^2$) defined in a geometric manner. We work with frames rather than coordinates because there is no geometric information at zero-th order.

For example, we could take a frame $\{\underline{\omega}^1, \underline{\omega}^2\}$, $\underline{\omega}^j \in \Omega^1(\mathbb{R}^2)$, such that

(a) $\underline{\omega}^1$ annihilates $L_1$,

(b) $\underline{\omega}^2$ annihilates $L_2$,

(c) $\underline{\omega}^1 - \underline{\omega}^2$ annihilates $L_3$.

For example, in the case of an ODE in coordinates, we could take $\underline{\omega}^1 = F(x, y)dx, \underline{\omega}^2 = dy$.

*Remark.* Note that we are imitating the flat model on the infinitesimal level. Just as in Riemannian geometry the tangent space to a point of a Riemannian manifold looks like Euclidean space, any web looks flat to first order.

While such a framing is arrived at using geometric considerations, it is not unique. An essential point is to only utilize objects that are defined by the geometry of a situation, so each time we make a choice, we will work on a larger space that includes the indeterminacy as independent variables.

Any other frame satisfying (a) and (b) must satisfy

$$\tilde{\underline{\omega}}^1 = \lambda \underline{\omega}^1$$
$$\tilde{\underline{\omega}}^2 = \mu \underline{\omega}^2$$

for some functions $\lambda, \mu$. Any frame satisfying (c) must be of the form

$$\tilde{\underline{\omega}}^2 - \tilde{\underline{\omega}}^1 = \nu(\underline{\omega}^2 - \underline{\omega}^1).$$

Combining these three conditions, we see $\lambda = \mu = \nu$.

We see that our choice of frame is unique up to a nonvanishing function $\lambda$, and following the remark above, we introduce $\lambda$ as an independent variable:

Let $\mathcal{F}_{\mathcal{L}} \to \mathbb{R}^2$ be the space of frames satisfying (a,b,c). I.e., points of $\mathcal{F}_{\mathcal{L}} \to \mathbb{R}^2$ may be written $(p, \underline{\omega}^1, \underline{\omega}^2)$ where $p \in \mathbb{R}^2$ and $\underline{\omega}^1, \underline{\omega}^2$ is a coframing satisfying (a,b,c). $\mathcal{F}_{\mathcal{L}}$ has dimension three, with local coordinates $(x, y, \lambda)$.

Dually, if we write $f = (p, e_1, e_2)$, where $(e_1, e_2)$ is the dual basis to $\underline{\omega}^1, \underline{\omega}^2$, then $e_1$ is tangent to $L_{2,p}$, $e_2$ is tangent to $L_{1,p}$, and $e_1 + e_2$ is tangent to $L_{3,p}$.

On $\mathcal{F}_{\mathcal{L}}$ we have *tautological forms*

$$\begin{pmatrix} \omega^1 \\ \omega^2 \end{pmatrix} := \begin{pmatrix} \lambda & 0 \\ 0 & \lambda \end{pmatrix}^{-1} \begin{pmatrix} \underline{\omega}^1 \\ \underline{\omega}^2 \end{pmatrix}.$$

**Exercise.** $\omega^1, \omega^2$ are independent of our initial choice of $\underline{\omega}^1, \underline{\omega}^2$.

**Aside/Definition.** Given a fibration

$$F \xrightarrow{i} E$$
$$\downarrow \pi$$
$$B$$

$\phi \in \Omega^k(E)$ is said to be *semi-basic* if $i^*(\phi) = 0$, i.e., if $\phi$ is zero on directions tangent to the fibers, equivalently, if $v \lrcorner \phi = 0$ for all $v \in \ker \pi_*$. (Where $(v \lrcorner \phi) \in \Omega^{k-1}(E)$ is the natural contraction.) $\phi$ is said to be *basic* if $\phi = \pi^*(\underline{\phi})$ for some $\underline{\phi} \in \Omega^k(B)$, i.e., if $\phi$ is the pullback of some form on $B$.

If we let $\{x^i\}$ be local coordinates on $B$ and $\{x^i, y^\alpha\}$ be local coordinates on $E$, then the general 1-form on $E$ may be written as

$$\phi = f_j(x^i, y^\alpha)dx^j + g_\beta(x^i, y^\alpha)dy^\beta$$

and $\phi$ is semi-basic if $g_\beta \equiv 0$ and basic if, in addition, the $f_j$ are functions of $x^i$ alone.

**Exercise.** $\omega^1, \omega^2$ are semi-basic, but not basic.

We now seek a geometrically defined coframing of $\mathcal{F}_\mathcal{L}$. Fortunately we already have two 1-forms defined geometrically, but $\dim \mathcal{F}_\mathcal{L} = 3$ so we seek a third 1-form. Not knowing what to do next, we differentiate:

$$d \begin{pmatrix} \omega^1 \\ \omega^2 \end{pmatrix} = - \begin{pmatrix} \frac{d\lambda}{\lambda} & 0 \\ 0 & \frac{d\lambda}{\lambda} \end{pmatrix} \wedge \begin{pmatrix} \omega^1 \\ \omega^2 \end{pmatrix} + \lambda \begin{pmatrix} d\underline{\omega}^1 \\ d\underline{\omega}^2 \end{pmatrix}$$

Since $d\underline{\omega}^j$ is semi-basic, we may write $\lambda d\underline{\omega}^j = T^j \omega^1 \wedge \omega^2$ for some functions $T^1, T^2$. Write $\theta = \frac{d\lambda}{\lambda}$. Our equation has the form

(*)
$$d \begin{pmatrix} \omega^1 \\ \omega^2 \end{pmatrix} = - \begin{pmatrix} \theta & 0 \\ 0 & \theta \end{pmatrix} \wedge \begin{pmatrix} \omega^1 \\ \omega^2 \end{pmatrix} + \begin{pmatrix} T^1 \omega^1 \wedge \omega^2 \\ T^2 \omega^1 \wedge \omega^2 \end{pmatrix}$$

We will refer to the terms $T^1, T^2$ as the *torsion of* $\theta$. The forms $\omega^1, \omega^2, \theta$ give a coframing of $\mathcal{F}_\mathcal{L}$, but $\theta$ is not uniquely determined. In fact the choice of a third form such that the derivative of our tautological forms $\omega^1, \omega^2$ may be expressed as in (*) is unique up to translates by $\omega^1, \omega^2$, i.e., any other choice must be of the form $\tilde{\theta} = \theta + a\omega^1 + b\omega^2$. If we choose $\tilde{\theta} = \theta - T^2\omega^1 + T^1\omega^2$, our new choice has the effect that $\tilde{T}^j = 0$ and there is a unique such form. Renaming $\tilde{\theta}$ as $\theta$ we have:

**Proposition.** *There exists a unique form* $\theta \in \Omega^1(\mathcal{F}_\mathcal{L})$ *such that the equations*

$$d \begin{pmatrix} \omega^1 \\ \omega^2 \end{pmatrix} = - \begin{pmatrix} \theta & 0 \\ 0 & \theta \end{pmatrix} \wedge \begin{pmatrix} \omega^1 \\ \omega^2 \end{pmatrix}$$

*are satisfied.*

Any choice of $\theta$ such that (*) holds will be called a *connection* and a choice of connection such that the torsion is zero, a *torsion free connection*.

We now have a space with a geometrically defined coframing. What do we do next? Differentiate!

We already have calculated $d\omega^1, d\omega^2$. To calculate $d\theta$ we compute

$$0 = d^2\omega^1 = -d\theta \wedge \omega^1 - \theta \wedge (\theta \wedge \omega^1) = -(d\theta + \theta \wedge \theta) \wedge \omega^1$$
$$0 = d^2\omega^2 = -d\theta \wedge \omega^2 - \theta \wedge (\theta \wedge \omega^2) = -(d\theta + \theta \wedge \theta) \wedge \omega^2$$

If $\alpha \in \Omega^1(\mathcal{F}_\mathcal{L})$ is any form, then $d\alpha = A\theta \wedge \omega^1 + B\theta \wedge \omega^2 + C\omega^1 \wedge \omega^2$, for some functions $A, B, C$. Since $d\theta \wedge \omega^j = 0$, we see

$$d\theta = K\omega^1 \wedge \omega^2$$

for some function $K$.

**Exercise.** The function $K$ is basic. It is called the *curvature* of the web.

The function $K$ is a *differential invariant*. A necessary condition for two webs to be locally equivalent is that they have the same web-curvature functions.

**Exercise.** (The fun part.) Recall that the Gauss curvature of a surface $M^2 \subset \mathbb{E}^3$ with Gauss map $\gamma : M \to S^2$ can be interpreted in terms of limits of

$$\lim_{U \to x} \frac{\text{area}\,\gamma(U)}{\text{area}\,(U)}$$

where $U \subset M$ is an open neighborhood of $x$. Show that there is a similar interpretation of the web curvature $K$ in terms of a limiting quotient as follows: First observe that in the flat case, if one begins at $x$, travels along say $L_1$ a 'distance' $d$ to a point $y$, then from $y$ one travels a 'distance' $d$ along $L_2$, then a 'distance' $d$ along $L_3$, then successively along $L_1, L_2, L_3$ again, one traces out a closed 'hexagon' (which is literally a hexagon if the web is given by lines). If the web is not flat the figure will not close up (for $d$ sufficiently small). Now let $d$ get small and take a limit to arrive at $K$.

**Exercise.** Calculate the curvature of the following webs:
1. $\{x = const, y = const, x - Cy = const\}$ where $C$ is a constant.
2. $\{x = const, y = const, \frac{x}{y} = const\}$.
3. $\{x = const, y = const, x^2 + y^2 = const\}$.

## 2   Second example: Riemannian geometry

Let $(M^n, g)$ be an oriented Riemannian manifold, that is, a differentiable manifold endowed with a section $g \in \Gamma(M, S^2_+ T^* M)$ and an orientation. Fix $x \in M$, a natural choice of basis of $T^*_x M$ would be an oriented orthonormal basis.

Let $x^1, \cdots, x^n$ be local coordinates on $M$. There exist functions $h = (h^i_j(x))$ such that $\underline{\omega}^j = (h^{-1})^i_j dx^j$ is an oriented orthonormal framing, i.e., an oriented orthonormal basis of $T^*_x M$ varying smoothly with $x$. We often use matrix notation and write $\underline{\omega} = h^{-1} dx$. $\underline{\omega}$ is unique up to the choice

$$\begin{pmatrix} \tilde{\omega}^1 \\ \vdots \\ \tilde{\omega}^n \end{pmatrix} = a(x)^{-1} \begin{pmatrix} \omega^1 \\ \vdots \\ \omega^n \end{pmatrix}$$

with $a(x) \in SO(n)$. Let $\mathcal{F}_{on}$ denote the space of all oriented orthonormal frames. After choosing functions $h_j^i$, locally we may specify a point of $\mathcal{F}_{on}$ by a pair $(x, a)$. Define $\omega_{(x,a)} := a^{-1}\underline{\omega}_x$.

**Exercise.** $\omega$ is well defined, independent of our choices of coordinates and functions $h_j^i$. As in our first example, we now seek a geometric coframing of $\mathcal{F}_{on}$, so we need to find a set of forms complementary to the $\omega^i$. We calculate

$$d\omega = -a^{-1}da \wedge a^{-1}\underline{\omega}_x + a^{-1}d\underline{\omega}_x$$

In analogy with our first example, we would like to write

$$(*) \qquad\qquad d\omega^j = -\theta_k^j \wedge \omega^k + T_{kl}^j \omega^k \wedge \omega^l$$

where $\theta_k^j + \theta_j^k = 0$. We will call a matrix valued 1-form $\theta = (\theta_j^i)$ satisfying (*) a *connection*. For example, we could define $\theta_k^j = (a^{-1}da)_k^j$ and observe that $a^{-1}d\underline{\omega}_x$ is semi-basic and the forms $\omega^i \wedge \omega^j$ span the semi-basic two-forms. Our group of admissible changes of frame is $SO(n)$ and we are requiring $\theta = (\theta_j^i)$ to have values in its Lie algebra $\mathfrak{so}(n)$. (Looking back to our first example, we see that there we had required the connection to be a one form with values in the Lie algebra of the group of admissible changes of frame as well.)

Fix an identification $V \simeq \mathbb{R}^n$ with its standard basis. Without indices (*) reads

$$d\omega = -\theta \wedge \omega + T$$

where $\omega \in \Omega^1(\mathcal{F}_{on}, V)$, $\theta \in \Omega^1(\mathcal{F}_{on}, \mathfrak{so}(V))$ and $T \in \Omega^2(\mathcal{F}_{on}, V)$.

The forms $\theta_j^i$ that appear in (*) are not canonically defined. To improve the situation, we seek a connection $\tilde{\theta}$ such that the torsion of the connection $\tilde{\theta}$ is zero. Any connection $\tilde{\theta}$ must be skew symmetric, so the most general change of connection is

$$\tilde{\theta}_j^i = \theta_j^i + A_{jk}^i \omega^k$$

with $A_{jk}^i + A_{ik}^j = 0$ for all $i, j, k$. In other words we must have $A = A_{jk}^i v_i \otimes v^j \otimes v^k \in \mathfrak{so}(V) \otimes V^*$ to insure that $\tilde{\theta} \in \Omega^1(\mathcal{F}_{on}, \mathfrak{so}(V))$. Note that we may assume that $A_{jk}^i = -A_{kj}^i$ as any three tensor skew in two indices and symmetric in the other two must be zero. (In fancy language, $A_{jk}^i = -A_{kj}^i$ because $\mathfrak{so}(V)^{(1)} := (\mathfrak{so}(V) \otimes V^*) \cap (V \otimes S^2 V^*) = 0$.)

If we choose

$$A_{jk}^i = \frac{1}{2}(T_{jk}^i - T_{ik}^j)$$

then $\tilde{T}^i_{jk} = 0$. (In fancy language, the skew symmetrization map $\delta : \mathfrak{so}(V) \otimes V^* \to V \otimes \Lambda^2 V^*$ is surjective, so $H^{02}(\mathfrak{so}(V)) := (V^* \otimes \Lambda^2 V)/\delta(\mathfrak{so}(V) \otimes V) = 0$.) We have proven:

**The fundamental lemma of Riemannian geometry.** *There exists a unique torsion free connection on* $\mathcal{F}_{on}(M)$.

To see the relation between this notion of a connection and others you may be more familiar with, see §5.

Now that we have a canonical coframing, any derivatives we calculate will have geometric meaning. We have already computed $d\omega$ so we compute $d\theta$. To do this we compute $0 = d^2\omega$:

$$0 = d^2\omega = (d\theta + \theta \wedge \theta) \wedge \omega$$

**Exercise.**

1. $\tilde{R} := d\theta + \theta \wedge \theta$ is semi-basic. In fact $\tilde{R}$ is basic and therefore descends to give an element $R \in \Omega^2(M, \mathfrak{so}(T_x M))$. $R$ is called the *Riemann curvature tensor*. One often writes $d\theta^i_j + \theta^i_p \wedge \theta^p_j = R^i_{jkl}\omega^k \wedge \omega^l$.

2. Prove the *first Bianchi identity*, $R^i_{jkl} + R^i_{klj} + R^i_{lkj} = 0$.

## 3 Third example: path geometry

### 3.1 First formulation of the problem

Given an ODE it would be useful to know if it is a familiar ODE in disguise.

Even better would be to have a classification of ODE's up to some notion of 'equivalence'. We already have seen one notion of equivalence for first order ODE's in the case of 3-webs in $\mathbb{R}^2$. The $G$-structure we now study will enable us to classify second order ODE's in $\mathbb{R}^2$ under the following notion of equivalence:

Given two second order ODE's in $\mathbb{R}^2$,

$$y'' = F(x,y,y') \quad y'' = G(x,y,y')$$

when does there exist a diffeomorphism $\Phi : \mathbb{R}^2 \to \mathbb{R}^2$ taking integral curves of $F$ to integral curves of $G$?

Note that such a notion of equivalence is not useful for first order ODE's because under an arbitrary diffeomorphism, all such are equivalent.

*3.2 Second formulation of the problem*

Let $X$ be a surface. A *set of paths on* $X$ will be a set of curves $C$ on $X$ such that for all $x \in X$ and for all $l \in \mathbb{P}T_xX$, there is a unique $c \in C$ such that $x \in c$ and $\mathbb{P}T_xc = l$.

**Example.** $X$ with a Riemannian or more generally Finsler metric.

**Example.** The paths consisting of circles of radius one in $\mathbb{R}^2$.

**Example.** Let $X = \mathbb{RP}^2$ have the set of paths given by linear $\mathbb{P}^1$'s. We call this example the *flat set of paths*.

**Question.** Which Riemannian metrics induce the flat set of paths?

It is clear that the flat metric induces the flat set of paths because $\mathbb{R}^2$ embeds as an open subset of $\mathbb{RP}^2$.

**Proposition.** The sphere $S^2$ with its standard metric induces the flat set of paths.

*Proof.* We need to show there is a local diffeomorphism $S^2 \to \mathbb{RP}^2$ that takes geodesics on the sphere to lines. Consider $S^2 \subset \mathbb{R}^3 \subset \mathbb{RP}^3$. Let $\mathbb{RP}^2_\infty = \mathbb{RP}^3 \backslash \mathbb{R}^3$. The geodesics on $S^2$ are exactly the intersection of $S^2$ with planes passing through the origin in $\mathbb{R}^3$ which can be extended uniquely to $\mathbb{RP}^2$'s. But the paths in $\mathbb{RP}^2_\infty$ are exactly the intersections $\mathbb{RP}^2_\infty \cap L$ where $L$ is a $\mathbb{P}^2$. We define the map from the hemisphere to $\mathbb{RP}^2_\infty$ using lines emanating from the origin and it is clear that geodesics on the sphere map to open subsets of linear $\mathbb{RP}^1$'s.

**Exercise.** Prove that the paths of geodesics on hyperbolic space $H^2$ is the flat set of paths.

*Remark.* The Beltrami theorem asserts that the only Riemannian metrics inducing the flat set of paths are those of constant sectional curvature.

Unfortunately, the problem of determining equivalence of sets of paths cannot be set up as a first order $G$-structure problem on $X$, as one needs second order data to specify a path geometry. We will thus work on a larger space.

A path is determined by a point $p \in X$ and a tangent direction $[v] \in \mathbb{P}T_pX$. Consider

$$\mathcal{P} := \mathbb{P}(TX)$$
$$\downarrow$$
$$X$$

we will set up the problem of equivalence of sets of paths as a $G$-structure problem on $\mathbb{CP}$.

Before defining the $G$-structure, we make an observation:

Lifting the paths in $X$ to $\mathcal{P}$ gives a 2 parameter family of curves in $\mathcal{P}$ which foliates $\mathcal{P}$. Call the leaves of this foliation $L_1$. Notice that the fibers of the fibration

$$S^1 = \mathbb{RP}^1 \longrightarrow \mathcal{P}$$
$$\downarrow \pi$$
$$X$$

induce another foliation of $\mathcal{P}$ by the fibres of $\pi$. Call the leaves of this foliation $L_2$.

We have the following picture:

$$\mathcal{P}$$
$$\swarrow \quad \searrow$$
$$Z^2 = \mathcal{P}/L_1 \quad X^2 = \mathcal{P}/L_2.$$

$Z$ is the space of paths in $X$. If the path geometry results from a second order ODE in the plane, we may think of $Z$ as the space of initial conditions for the ODE.

While a point in $Z$ corresponds to a path in $X$, a point $x \in X$ also determines a curve in $Z$ (the curve of paths passing through $x$). The transform of $X$ induces a set of paths in $Z$, called the *dual set of paths*. $\mathcal{P}$ may be thought of as an incidence correspondence:

$$\mathcal{P} = \{(z,x) \mid x \in \text{ some path in } X \text{ corresponding to } z\}$$
$$= \{(z,x) \mid z \in \text{ some path in } Z \text{ corresponding to } x\}$$

The following example will turn out to be our "flat" model:

Let $X = \mathbb{RP}^2$ have the standard set of paths. Then $Z = \mathbb{P}^{2*} = \{\text{hyperplanes in } \mathbb{P}^2\}$, $\mathcal{P} = \{(x,H) \mid x \in H\}$, the incidence correspondence. $X, Z, \mathcal{P}$ are all homogeneous spaces of $SL(3, \mathbb{R})$. The paths in $\mathbb{P}^2$ are the linear $\mathbb{P}^1$'s, and the paths in $\mathbb{P}^{2*}$ are the linear $\mathbb{P}^{1*}$'s.

*Remark.* In dimensions greater than two, the space of paths is higher dimensional than the original space. For example, if $X = \mathbb{P}^n$, then $Z = G(\mathbb{P}^1, \mathbb{P}^n) = G(2, n+1)$. Here $\mathcal{P}$ is the space of flags, $\hat{x} \subset \hat{l} \subset V$.

Of particular interest to physicists is the complex case $X = \mathbb{CP}^3$, $Z = G(\mathbb{C}^2, \mathbb{C}^4)$ which relates to complexified compactified Minkowski space.

### 3.3 Setting up the G-structure

We define a *path geometry* on a 3-manifold $M$ to be the following data: two line bundles $L_1, L_2 \subset TM$ such that
1. $L_1 \cap L_2 = 0$
2. The 2-plane bundle $\{L_1, L_2\}$ is nowhere integrable.

**Example.** Let $(x, y, p)$ be coordinates on $J^1(\mathbb{R}, \mathbb{R})$, the space of one-jets of mappings $\mathbb{R} \to \mathbb{R}$. (If $f : \mathbb{R} \to \mathbb{R}$ is a mapping, it has a canonical lift to $J^1$ given by $(x = x, y = f(x), p = f'(x))$.) One has

$$L_1 = \ker\{dx, dy - pdx\}$$
$$L_2 = \ker\{dp - Fdx, dy - pdx\}$$

and $\mathrm{span}\{L_1, L_2\}$ determines the contact system (it is annihilated by the contact form $dy - pdx$).

Let $\mathbb{R}^3$ have its standard basis $e_1, e_2, e_3$ and let $G \subset Gl_3\mathbb{R}$ be the subgroup preserving $\{e_1\}$ and $\{e_3\}$. i.e.,

$$G = \left\{ \begin{pmatrix} a & e & 0 \\ 0 & b & 0 \\ 0 & f & c \end{pmatrix} \subset Gl_3\mathbb{R} \right\}.$$

Let $M^3$ be equipped with a path geometry and define $\mathcal{F}_G \to M$ by

$$\mathcal{F}_G := \{u \in \mathcal{F}_{Gl(V)} \mid u(L_1) = \{e_1\},\ u(L_2) = \{e_3\}\}.$$

Let $\omega = \begin{pmatrix} \omega^1 \\ \omega^2 \\ \omega^3 \end{pmatrix}$ be the *tautological* $\mathbb{R}^3$-valued 1-form on $\mathcal{F}_G$, that is, we take any section $\underline{\omega}$ of $\mathcal{F}_G$ in coordinates and define $\omega_{(x,a)} := a^{-1}\underline{\omega}$ (where $x \in M, a \in G$. (See §4 below for the precise definition). We have arranged things such that $\{\omega^2, \omega^3\} = \pi^*(L_1{}^\perp)$ and $\{\omega^1, \omega^2\} = \pi^*(L_2{}^\perp)$. In particular, the pullback of $\omega^2$ along any section is a contact form.

Let $\theta$ be any choice of connection, that is a $\mathfrak{g}$ valued one-form (where $\mathfrak{g}$ is the Lie algebra of $G$) that enables us to write:

$$d\omega = - \begin{pmatrix} \theta_1^1 & \theta_1^2 & 0 \\ 0 & \theta_2^2 & 0 \\ 0 & \theta_2^3 & \theta_3^3 \end{pmatrix} \wedge \begin{pmatrix} \omega^1 \\ \omega^2 \\ \omega^3 \end{pmatrix} + \begin{pmatrix} T_{ij}^1 \omega^i \wedge \omega^j \\ T_{ij}^2 \omega^i \wedge \omega^j \\ T_{ij}^3 \omega^i \wedge \omega^j \end{pmatrix}$$

By choosing a new connection $\tilde{\theta}$ we can arrange for all torsion terms except for $T_{13}^2 \omega^1 \wedge \omega^3$ to be zero. The $G$-structures arising from path geometries will never have $T_{13}^2 = 0$ because $\omega^2$ pulls back to be a contact form. Thus the situation here is different from our first two examples. We need to learn how to deal with torsion. We first use our freedom to simplify it. If $g \in G$, then (exercise) $g(T_{13}^2) = \frac{b}{ac}T_{13}^2$. We thus can and do restrict to frames where $T_{13}^2 \equiv 1$. A further calculation shows that, unlike our previous cases, the choice of connection preserving our normalized torsion is not unique. Rather than make a choice with no geometric meaning, we work on a larger space, call it $\mathcal{F}_{G_1}{}^{(1)}$, where the ambiguity in choice of connection is introduced as independent variables, just as we did in the outset with the ambiguity in the choice of adapted frame. Fortunately on this new space there does exist a torsion free connection and we can make it unique by reducing our group. Call the new group $G_2$. Since we have a unique connection, we have a canonical coframing of $\mathcal{F}_{G_2}$ and we take derivatives to find differential invariants. There are two functionally independent relative differential invariants, call them $H_1, H_2$. (From these one can construct tensors that are invariants, i.e., tensors that are basic for the projection to $M$. In particular there is a cubic differential $\Psi$ with coefficient $H_2$.) For more details, see [1] where the calculations are carried out explicitly.

Finally the fun part: we interpret the invariants geometrically. In particular, we verify that if $H_1 = H_2 = 0$, the $G$-structure comes from the flat set of paths. (Note a slight anomaly of terminology- the flat set of paths does not induce a flat $G$-structure because there was torsion.)

Consider first the case of paths arising from an ODE in the plane. If we choose frames as above, we obtain:

$$H_2 = F_{pppp}$$
$$H_1 = D_x^2 F_{pp} - 4D_x F_{py} + F_p(4F_{py} - D_x F_{pp}) + 3F_y F_{pp} + 6F_{yy}$$

where $D_x(g(x,y,p)) = g_x + g_y \frac{\partial y}{\partial x} + g_p \frac{\partial p}{\partial x}$.

We see that $H_2 = 0$ implies our system of paths is locally equivalent to the integral curves of an ODE of the form

$$(**) \qquad\qquad y'' = A(y')^3 + B(y')^2 + C(y') + D$$

where $A, B, C, D$ are functions of $x, y$. (Such a set of paths gives rise to a *normal projective connection*, a consequence of this is that $H_1$ descends to become a differential invariant of the set of paths.) Now for the case $H_1 = 0$. While this condition does not have any obvious consequences for our original path geometry, by duality, we know that the dual system of paths is equivalent to the integral curves of an ODE of the form (**). If $H_1 = 0$ and $H_2 \neq 0$, then, except for a few exceptions (see [1] and the references therein), the equation $y'' = F(x,y,p)$ is solvable by quadrature. What happens here is that $H_2$

descends to an invariant on $M/L_2$ which is constant along integral curves, i.e., it determines a first integral.

*Remark.* If we work over $\mathbb{C}$ and in the holomorphic category and we require the integral curves in $Z$ to be $\mathbb{CP}^1$'s, then since $\mathbb{CP}^1$ does not support any holomorphic cubic differential, $\Psi \equiv 0$ and thus $H_2 \equiv 0$. This is most pleasantly exploited by Hitchin in [2].

## 4 Definitions

We fix $V = \mathbb{R}^n$ with its standard basis. Over any differentiable manifold $M^n$ we define the general frame bundle $\pi : \mathcal{F}_{Gl(V)} \to M$ whose points $f = (x, u)$ are linear isomorphisms $u : T_x M \to V$. $\mathcal{F}_{Gl(V)}$ comes equipped with a tautological $V$-valued one-form $\omega$ defined by $\omega_{(x,u)}(w) = u(\pi_*(w))$ and thus a canonical basis $\omega^i$ of the semi-basic forms by writing $\omega := \omega^i v_i$ where $v_i$ is the standard basis of $V$.

**Definition.** Let $G \subseteq Gl(V)$ be a matrix Lie group. A *(first-order) G-structure on $M$* is a smoothly varying $G$-module structure on $T_x M$. In other words, for all $x \in M$ there exists a map $\rho_x : G \to Gl(T_x M)$. We let $\mathcal{F}_G \subset \mathcal{F}_{Gl(V)}$ be the subbundle of $G$-frames, that is

$$\mathcal{F}_G := \{(x, u) \in \mathcal{F}_{Gl(V)} \mid u : T_x M \to V \text{ is } G\text{-equivariant} \}$$
$$= \{(x, u) \in \mathcal{F}_{Gl(V)} \mid u(\rho_x(g)v) = g(u(v)) \ \forall v \in T_x M, \ \forall g \in G\}$$

$G$ is often described as the group preserving some additional structure on $V$, e.g. $O(V) = O(V, <>)$, the group preserving an inner product $<,> \in S^2 V^*$, $Gl(m, \mathbb{C}) = Gl(V, J) \subset Gl(V)$ as the group of elements commuting with an almost complex structure $J$, $Sp(V, \phi)$ as the group preserving a skew-form $\phi \in \Lambda^2 V^*$. If $G$ is defined by a form $\psi \in \mathbb{T}(V)$, where $\mathbb{T}(V)$ is some tensor space (e.g., $\mathbb{T} = S^2 V, \Lambda^2 V, V^* \otimes V$) then specifying a $G$-structure is equivalent to specifying a smooth section $\psi \in \Gamma(M, \mathbb{T}(TM))$. If $u : T \to V$ is a linear map, let $u_{\mathbb{T}} : \mathbb{T}(T) \to \mathbb{T}(V)$ be the induced linear map. Given $G \subset Gl(V)$, let $\rho_{\mathbb{T}} : G \to Gl(\mathbb{T}(V))$ be the induced representation, then $\mathcal{F}_G = \{(x, u) \in \mathcal{F}_{GL(V)} \mid u_{\mathbb{T}}^*(\psi) = \psi\}$.

$G$ acts on $\mathcal{F}_{Gl(V)}$ by $g(x, u) = (x, ug)$ and we may form the quotient bundle $\mathcal{F}_{Gl(V)}/G$.

**Exercise.** A (first order) $G$-structure on $M$ is equivalent to specifying a smooth section $s : M \to \mathcal{F}_{Gl(V)}/G$.

**Proposition/Definition.** *Two $G$-structures $s, s'$ are equivalent if any (and therefore all) of the following hold:*

i.  *there exists a local diffeomorphism $\phi : M \to M$ such that $\phi^*(s') = s$.*

ii. *there exists a local diffeomorphism $\phi : M \to M$ such that $\phi_{*x} : T_x M \to T_{\phi(x)}M$ is an isomorphism of $G$-modules, where $T_x M$ has the $s$-$G$-module structure and $T_{\phi(x)}M$ has the $s'$-$G$-module structure.*

iii. *there exists a local diffeomorphism $\Phi : \mathcal{F}_G \to \mathcal{F}_{G'}$ such that $\Phi^*(\omega') = \omega$.*

**Exercise.** Prove the proposition.

**Definition.** A $G$ structure is *flat* if there exists local coordinates $x^1, \cdots, x^n$ on $M$ such that $(dx^1, \cdots, dx^n)$ is a section of $\mathcal{F}_G$, i.e., such that $(dx^1, \cdots, dx^n)$ is a $G$-framing.

For example, an $O(n)$-structure (Riemannian metric) is flat iff there exist local coordinates $x^1, \cdots, x^n$ such that $dx^i$ gives a local orthonormal framing of $M$.

To find invariants of a $G$-structure, we find a geometrically defined framing of $\mathcal{F}_G$ and then take derivatives. We already have a basis of the semi-basic forms and the fiber $\mathcal{F}_{G,x}$ is diffeomorphic to $G$. Any Lie group $G$ has a canonical framing by its *Maurer-Cartan form*, $\underline{\theta} \in \Omega^1(G, \mathfrak{g})$. Where, if $a \in G$, we define $\underline{\theta}_a = a^{-1}da$. (Exercise: verify that indeed $\underline{\theta}$ is $\mathfrak{g}$-valued and obeys the *Maurer-Cartan equation* $d\underline{\theta} = -\underline{\theta} \wedge \underline{\theta}$.)

**Definition.** A *connection* $\theta$ on $\mathcal{F}_G$ is a $\mathfrak{g}$-valued one-form, $\theta \in \Omega^1(\mathcal{F}_G, \mathfrak{g})$ satisfying:

$$d\omega = -\theta \wedge \omega + T$$

where $T \in \Omega^2(\mathcal{F}, V)$ is semi-basic. In indices

$$d\omega^i = -\theta^i_j \wedge \omega^j + T^i_{jk}\omega^j \wedge \omega^k$$

where $\theta^i_j v_i \otimes v^j \in \Omega^1(\mathcal{F}_G, \mathfrak{g})$ and $T^i_{jk} = -T^i_{kj}$.

If we fix a point $u_0$ in a fiber $\mathcal{F}_{G,x}$, then there is a mapping $\mu_{u_0} : G \to \mathcal{F}_{G,x}$ defined by $g \mapsto u_0 g$.

**Proposition.** Let $\underline{\theta} \in \Omega^1(G, \mathfrak{g})$ denote the Maurer-Cartan form of $G$. A $\mathfrak{g}$-valued 1-form $\theta$ on $\mathcal{F}_G$ is a connection form if and only if for all $x \in M$ and for any $\mu_{u_0} : G \to \mathcal{F}_{G,x}$

   1. $\mu_{u_0}^*(\theta) = \underline{\theta}$

and

   2. $T^*\mathcal{F}_G$ is spanned by the entries of $\theta$ and $\omega$.

**Exercise.** Show that if $u_1 \in \mathcal{F}_{G,x}$ is another element, that $\mu_{u_0}^*(\theta) = \mu_{u_1}^*(\theta)$.

Note that 2. is equivalent to requiring $\ker \theta \cap \ker \pi_* = 0$, as $\dim \mathcal{F}_G = \dim G + \dim M$.

*Proof.* Locally $\omega = a^{-1}h^{-1}dx$, where $h : M \to Gl(V)$ is such that $\underline{\omega} = h^{-1}dx \in \Gamma(M, \mathcal{F}_G)$ and $a \in G$. We calculate

$$d\omega = -a^{-1}daa^{-1}h^{-1} \wedge dx - ah^{-1}dhh^{-1} \wedge dx$$
$$= -(a^{-1}da) \wedge \omega - ah^{-1}dhh^{-1} \wedge dx$$

The second term is semi-basic, thus it can be written as a linear combination of the $\omega^i \wedge \omega^j$. $\mu_{u_0}^*(a^{-1}da) = \underline{\theta}$ so $a^{-1}da$ is a choice of connection. Any other choice of connection would be of the form $a^{-1}da + \beta$ where $\beta$ is semi-basic (and $\mathfrak{g}$-valued) and thus 1. and 2. are not affected.

**Definition.** $T := T_{jk}^i v_i \otimes v^j \wedge v^k \in \Gamma(\mathcal{F}_G, V \otimes \Lambda^2 V^*)$ is called the *torsion of the connection* $\theta$. If we change our connection, we may be able to eliminate the torsion as before.

Let $H^{02}(\mathfrak{g}) = (V \otimes \Lambda^2 V^*)/\delta(\mathfrak{g} \otimes V^*)$ where $\delta : (V \otimes V^*) \otimes V^* \to V \otimes \Lambda^2 V^*$ is the skew symmetrization map.

**Definition.** Let $s$ be a $G$-structure on $M$, $\theta$ a connection form and let $T \in \Gamma(\mathcal{F}_G, V \otimes \Lambda^2 V^*)$ be the torsion of $\theta$. $[T] \in \Gamma(\mathcal{F}_G, H^{02}(\mathfrak{g}))$ is called the *torsion of the G-structure*.

**Exercise.** Show that $[T]$ is basic, i.e., it descends to a section $[T] \in \Gamma(M, H^{02}(\rho(\mathfrak{g})))$.

To determine the unicity of a (e.g. torsion free) connection, we need to calculate the changes of connection that do not alter the torsion. This is $\mathfrak{g}^{(1)} := \ker \delta = (\mathfrak{g} \otimes V^*) \cap (V \otimes S^2 V^*)$.

In the case $s$ is a $G$-structure with $[T] = 0$ and $\mathfrak{g}^{(1)} = 0$, we calculate $0 = d^2\omega = (d\theta + \theta \wedge \theta) \wedge \omega$ and observe that $R := d\theta + \theta \wedge \theta \in \Omega^2(\mathcal{F}_G, \mathfrak{g})$ is a well defined differential invariant, called the *curvature* of the $G$-structure. As in the case of Riemannian geometry, $R$ is basic and descends to an element $R \in \Gamma(M, \rho(\mathfrak{g}))$. If $\mathfrak{g}^{(1)} \neq 0$ and $\theta$ is a torsion free connection, then we may still define $R_\theta$, but now it is just the curvature of $\theta$.

**Exercise.** Show that the curvature of a $G$-structure with $\mathfrak{g}^{(1)} = 0$ is basic, i.e., $R$ descends to a well defined section $R \in \Omega^2(M, \rho(\mathfrak{g}))$

**Exercise (The first Bianchi identity)** Show that $R_x \in (\mathfrak{g} \otimes \Lambda^2 V^*) \cap (V \otimes S_{21} V^*) = \ker \delta_2$, where $\delta_2 : \mathfrak{g} \otimes \Lambda^2 V^* \to V \otimes \Lambda^3 V^*$ is the skew symmetrization map.

**Exercise.** Show that if $s$ is a $G$-structure on $M$ such that $\mathfrak{g}^{(1)} = 0$, $R = 0$, and $[T] = 0$, then $s$ is flat.

## What to do if $\mathfrak{g}^{(1)} \neq 0$.

Intuitively we know what to do here: we eliminate the ambiguity by working on a larger space, the enlargement eliminates making a choice of connection.

Let $\mathcal{F}_G{}^{(1)} \to \mathcal{F}_G$ be the principal bundle whose fiber is the additive group $\mathfrak{g}^{(1)}$. We now continue to differentiate on $\mathcal{F}_G{}^{(1)}$.

In this case (assuming $T_\theta = 0$), we can still define the curvature of the $G$-structure $R = [R_\theta]$ as an element of

$$H^{12}(\mathfrak{g}) := \frac{\ker \delta_2}{\text{Image } \delta_1}.$$

## What to do if $H^{02}(\mathfrak{g}) \neq 0$ (or, more precisely $[T] \neq 0$)

Intuitively, it is clear what to do here as well. We calculate how the torsion varies as one moves in the fiber, and choose some normalization of the torsion. This normalization of torsion will reduce the group of admissible changes to a subgroup $G_1 \subset G$ and we begin again, studying $G_1$-frames (as we did in our third example).

Problems can occur if the $G$ action on the torsion is not well behaved, so at this point we refer the reader to [3].

**Example: Almost complex manifolds.** Consider $Gl(m, \mathbb{C}) \subset \mathfrak{Gl}(2m, \mathbb{R})$ defined as the group preserving

$$J = \begin{pmatrix} 0 & -Id \\ Id & 0 \end{pmatrix}$$

We have

$$\mathfrak{g} = \mathfrak{gl}(m, \mathbb{C}) = \begin{pmatrix} A & -B \\ B & A \end{pmatrix}$$

where $A, B \in \mathfrak{gl}(m, \mathbb{R})$ are arbitrary. Note that $\dim \mathfrak{g} = 2m^2$. Let $1 \leq i, j, k \leq m$. We write $\{v_i \otimes v^j + v_{m+i} \otimes v^{m+j}, v_i \otimes v^{m+j} - v_{m+i} \otimes v^j\}$ as a basis of $\mathfrak{g}$. Consider

(1) $\quad \delta(v_i \otimes v^j + v_{m+i} \otimes v^{m+j}) \otimes v^l = v_i \otimes v^j \wedge v^l + v_{m+i} \otimes v^{m+j} \wedge v^l$

(2) $\delta(v_i \otimes v^j + v_{m+i} \otimes v^{m+j}) \otimes v^{m+l} = v_i \otimes v^j \wedge v^{m+l} + v_{m+i} \otimes v^{m+j} \wedge v^{m+l}$

(3) $\quad \delta(v_i \otimes v^{m+j} - v_{m+i} \otimes v^j) \otimes v^l = v_i \otimes v^{m+j} \wedge v^l - v_{m+i} \otimes v^j \wedge v^l$

(4) $\delta(v_i \otimes v^{m+j} - v_{m+i} \otimes v^j) \otimes v^{m+l} = v_i \otimes v^{m+j} \wedge v^{m+l} - v_{m+i} \otimes v^j \wedge v^{m+l}$

For fixed indices $i, j < l$, $(1) - (2) + (3) - (4) \in \ker \delta$. Thus $\dim \ker \delta \geq m\binom{m}{2}$, and in fact (exercise) $\dim \ker \delta = m\binom{m}{2}$. Thus for $m \geq 2$ it is possible for an $\mathfrak{gl}(m, \mathbb{C})$ structure to have torsion, but not for $m = 1$.

**Exercises.**

1. Let $s$ be a torsion free $\mathfrak{gl}(m, \mathbb{C})$ structure. Then $s$ is flat.

2. An $Sp(m, \mathbb{R}) \subset Gl(2m, \mathbb{R})$ structure is determined by a choice of $\phi \in \Omega^2(M)$ such that $\phi^{\wedge m} \neq 0$. Show that a torsion-free $Sp(m, \mathbb{R})$ structure is flat.

## 5  Relations with other notions of a connection

**Definition.** A *connection* on a fiber bundle $\pi : F \to M^n$ is a horizontal distribution $\Delta$, i.e., $\dim \Delta_x = n$ and $TF = \{\Delta, \ker \pi_*\}$.

Given a $G$-structure with bundle $\mathcal{F}_G$ (or more generally, any principal $G$-bundle over $M$), if $\mu : G \to Gl(W)$ is any representation of $G$, we can construct an induced vector bundle $E_\mu \to M$ by

$$E_\mu := \mathcal{F}_G \times_G W$$

where $(u, w) \in \mathcal{F}_G \times W$ is subject to the equivalence relation $(u, w) \simeq (ug, \mu(g^{-1})w)$.

### 5.1  Exercises.

1. The choice of a connection on $\mathcal{F}_G$ is equivalent to the choice of a horizontal distribution $\Delta \subset T\mathcal{F}_G$. Given $\theta$, one has the distribution $\Delta := \ker \theta$. Show that conversely $\theta$ is uniquely determined by its kernel.

2. If $\rho : G \to Gl(V)$ defines $G$, then $TM = E_\rho$ and $T^*M = E_{\rho^*}$.

3. Show that a connection on $\mathcal{F}_G$ induces connections on all the $E_\mu$ by, letting $pr : \mathcal{F}_G \times W \to E_\mu$ and defining $\Delta_{[(u,w)]} = pr_*(\Delta_u, T_w W)$.

4. Show that a connection on $\mathcal{F}_G$ gives rise to a differential operator $\nabla :$ $\Gamma(M, E_\mu) \to \Gamma(M, E_\mu \otimes T^*M)$ i.e., given $X \in T_x M$, $\nabla_X : \Gamma(M, E_\mu) \to \Gamma(M, E_\mu)$. Define $\nabla$ by $\nabla_X(s) = proj_{vert}(ds(X))$. Here $proj_{vert}\alpha$ is the vertical component of $\alpha$, and since $E$ is a vector bundle, there is a canonical isomorphism $T^{vert}E \simeq E$. Conversely, show that a first order differential operator $\nabla$ gives rise to a horizontal distribution on $TM$ and thus on $\mathcal{F}_G$, and is therefore equivalent to the choice of $\theta$.

5. Show that if $\nabla$ is the induced operator on $TM$, then for all $X, Y \in T_x M$, the maps

   i. $T(X, Y) := \nabla_X Y - \nabla_Y X - [X, Y] \in \Gamma(TM)$
   and

   ii. $R(X, Y) := \nabla_X \nabla_Y - \nabla_Y \nabla_X - \nabla_{[X,Y]} : \Gamma(TM) \to \Gamma(TM)$

are well defined. Identify $T$ with the torsion of $\theta$ and $R$ with the curvature of $\theta$.

## References

1. R. BRYANT, P.A. GRIFFITHS & L. HSU, *Towards a geometry of differential equations*, Geometry, Topology, & Physics, Conf. Proc. Lect. Notes. Geom. Top. VI, Internat. Press, Cambridge, MA, (1995), 1–76.
2. N. HITCHIN, *Complex manifolds and Einstein's equations*, LNM 970, Springer-Verlag, Berlin, (1982), 73–99.
3. J. M. LANDSBERG, *Cartan for beginners: an introduction to moving frames and exterior differential systems*, book in preparation

# WEB GEOMETRY AND THE EQUIVALENCE PROBLEM OF THE FIRST ORDER PARTIAL DIFFERENTIAL EQUATIONS

ISAO NAKAI

*Department of Mathematics, Faculty of Science, Ochanumizu University, Bunkyo-ku, Ontsuka 2-1-1 Tokyo 112-8610, Japan, nakai@math.ocha.ac.jp*

## Introduction

In this note the author will introduce Web geometry of solutions of holonomic first order partial differential equations (PDE) and also the theory of versal deformation of PDEs with first integrals. The purpose of this note is, as well as to develop a general theory of geometric invariants of PDEs, to seek geometric understanding of the structure of propagation of wave fronts, which is focused in the theory of optical caustics [5,36,62].

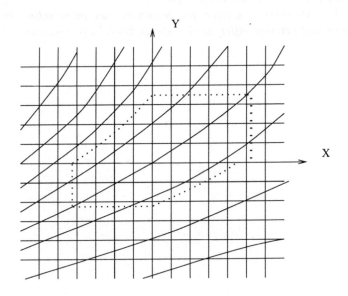

Figure 0.

The invariant theory of 3-webs on the plane had been studied by E. Cartan (see [14,64]) before it was named Web geometry by Blaschke and Bol [10].

Cartan initiated the classification theory (equivalence problem) for the ordinary differential equations in terms of connections on higher jet bundles. To explain the relation of his work to web geometry, consider an ordinary differential equation of one variable $dy/dx = f(x,y)$ on a neighbourhood of the origin. Clearly the solutions form a non singular foliation, which is defined by $dy - f\, dx = 0$. We say that two germs of ordinary differential equations $dy/dx = f(x,y), dy/dx = g(x,y)$ at $x = y = 0$ are *equivalent* if there exist germs of diffeomorphisms $\phi$ of $x$ and $\psi$ of $y$ respecting 0 such that $\phi \times \psi$ sends the foliation by the solutions of the former equation to that of the later equation. Clearly this equivalence preserves also the foliations defined by $dx = 0$ and $dy = 0$, hence the triple (3-web) of the foliations defined by $dy - f\, dx, dx, dy$ is sent to that defined by $dy - g\, dx, dx, dy$. In fig. 0 the 3-web structure is seen.

In this context Cartan defined an affine connection on the $xy$-space, which was later generalized and called the Chern connection. The connection form of this connection for the 3-web of $dy - f dx, dx, dy$ is $\theta I$, where $I$ is the identity $2 \times 2$ matrix and

$$d\theta = \frac{\partial^2}{\partial x \partial y} \log f\, dx \wedge dy,$$

which is called the web curvature form. After this observation, Cartan studied explicit differential equations of higher order

$$y^{(n)} = f(x, y, y', y^{(2)}, \ldots, y^{(n-1)})$$

and gave a "modern" insight on the works of Tresse (see [64]). This equation defines the one form $dy^{(n)} = f\, dx$ on the $n-1$-th order jet space $J^{n-1}(\mathbb{C}_x, \mathbb{C}_y)$ with the variables $(x, y, y', y^{(2)}, \ldots, y^{(n-1)})$, together with the higher order contact forms $dy^{(i-1)} - y^{(i)} dx, i = 1, 2, \ldots, n-2$ and $dx$. These forms form a coframe on the jet space, and the integrability condition of the system of these one forms gives a structure equation of a certain affine connection. The latest achievement on this problem is seen in the paper of Sato [64]. To explain the subject of this note, let us consider an implicit first order differential equation

$$f(x, y, y') = 0.$$

By solving $y'$ as an implicit function of $x, y$, one obtains many explicit differential equations

$$y' = f_1(x,y), \qquad y' = f_2(x,y), \qquad \ldots, \qquad y' = f_d(x,y)$$

and integrable one forms

$$dy - f_1\, dx, \qquad dy - f_2\, dx, \qquad \ldots, \qquad dy - f_d\, dx.$$

The solutions of these explicit differential equations form a configuration of $d$ foliations of codimension one at a generic point in $xy$-space. Such a structure is called a *solution d-web* or simply *d-web,* and in the non singular case where the foliations are all non singular and in general position, it was studied systematically by Blaschke and Bol [9],[10] and later by the Russian school [31],[32].

More generally a germ of *d-web of codimension 1* on $\mathbb{C}^n$ is a configuration $W = (F_1, \ldots, F_d)$ of $d$ foliations of codimension 1. By an easy formal calculation it is seen that the holomorphic equivalence classes of d-webs, $n + 1 \leq d$, form subsets of infinite codimension in the jet space of $d$-tuples of level functions defined at a $0 \in \mathbb{C}^n$. So analytic classification of the solution webs fails in the ordinary sense. In fact, Arnol'd [7], Carneiro [15], Dufour [23] and Hayakawa-Ishikawa-Izumiya-Yamaguchi [37] showed that analytic equivalence classes of some webs defined by PDEs with first integrals have moduli of infinite dimension, which are parameterized by the space of smooth functions defined on the configuration space the $\mathbb{C}^n$ at 0. And they called the parameter space the *function moduli.* The author [55] showed that for such web structure, topological equivalence is automatically given by a holomorphic diffeomorphism. So even topological classification does not make sense.

In this note we will generalize the above idea to define invariant forms for some cases and also we will investigate the relation of the singularities of the affine connection (the pole and zero of the connection form and the curvature form) to the singularities of the web structure (generalization of the branched locus of the implicit functions $f_1, \ldots, f_d$).

From now on a PDE on $\mathbb{C}_x^{n-1} \times \mathbb{C}_y$ is a subvariety $V \subset J^1(\mathbb{C}^{n-1}, \mathbb{C})$ of the first jet space of functions on $\mathbb{C}_x^{n-1}$ endowed with the contact 1-form $\omega = dy - \Sigma p_i dx_i$. The restriction of the contact form $\omega$ to $V$ has two kinds of singularities: the singularities of $V$, and the singularities of the restriction of $\omega$ to the smooth part of $V$. The second kind of singularity appears at those points where $V$ has contact with the contact elements, i.e. the planes defined by $\omega = 0$. We assume dim $V = n$ (in other words $V$ is a holonomic PDE) and also that the one form on the variety $V$ induced by the contact form is integrable at non singular points. By integrability, those integral manifolds of $\omega$ form a foliation of codimension 1 on the complement of the singular locus. Under the natural projection $\pi$ of $V$ onto the base space $\mathbb{C}^{n-1} \times \mathbb{C}$, those local integral manifolds in $V$ project to the *solutions* of the PDE. If the projection is $d$-to-one, the foliation on $V$ projects to the base space to form a $d$-configuration of foliations ($d$-web) of codimension 1 at a generic point. By the integrability,

$$dw = \theta \wedge \omega$$

holds locally on $V$ ($\theta$ being a 1-form). This is seen as a structure equation of Bott (partial) connection of the normal bundle of the foliation with coframe $\omega$. For simplicity assume that the projection is $(n+1)$-to-one. Then the foliation on $V$ projects to give an $(n+1)$-web on the base space.

Denote the direct image of the contact form $\omega$ on the $i$-th sheet of the projection of $V$ by $\omega_i$ for $i = 1, \ldots, n+1$. By integrability of $\omega_i$,

$$d\omega_i = \theta_i \wedge \omega_i$$

holds with one forms $\theta_i$ locally defined on the $i$-th sheet. The problem to define a canonical connection associated to a web is to seek canonical connection forms $\theta_i$. The method initiated by Blaschke and Goldberg [9,10,31] for local study of $(n+1)$-webs tells that there is a canonical choice of $\theta_i$ specializing the $i$-th $\omega_i$ and then the other equations for $j \neq i$ are written as a structure equation

$$d\begin{pmatrix} \omega_1 \\ \omega_2 \\ \cdots \\ \omega_{n+1} \end{pmatrix} = \begin{pmatrix} \theta_i & 0 & \cdots & 0 \\ 0 & \theta_i & \cdots & 0 \\ & & \cdots & \\ 0 & \cdots & 0 & \theta_i \end{pmatrix} \wedge \begin{pmatrix} \omega_1 \\ \omega_2 \\ \cdots \\ \omega_{n+1} \end{pmatrix} + T \qquad (*)$$

with coframe $\omega_1, \ldots, \hat{\omega}_i, \ldots, \omega_{n+1}$, torsion term $T$ and the connection form $\theta_i \cdot I$. To determine $\theta_i$ we impose a certain normalization condition on the torsion part in sect. 14. We will show that the resulting connection $\nabla_i$ depends on $i$ but is well defined on the $i$-th sheet of $V$. The connection $\nabla_i$ on each $i$-th sheet glue together to give a connection $\nabla$ on the smooth part of $V$. The connection on $V$ projects to the base space to define $(n+1)$ affine connections, which are also denoted by $\nabla_i, i = 1, \ldots, n+1$. It is shown that the mean of those connections i.e. the direct image $\pi_* d\theta$ has the curvature form equal to a constant multiple of the curvature form defined by Blaschke in 1930's [9]. The idea of mean connection enables us to define the connection $\nabla$ for all the cases $d \geq n+1$ (see sect. 16).

In the case $d \leq n$, $\omega_1 \wedge \cdots \wedge \omega_d$ defines a codimension $d$ foliation. A similar argument may apply to define a unique connection of the normal bundle of rank $d$. The case $d = n$ should be the simplest case.

In general Blaschke curvature form for a singular $(n+1)$-web is a rational 2-form on the base space singular along the set of those $(x, y)$ where the web structure is not generic. We will observe by some examples of 3-webs that the singularity locus of the curvature form reflects some geometric structure of the singularity locus of the web.

# 1 Introduction from the view point of the Stationary Phase Method

The study of the oscillatory integral

$$I_\lambda(q) = \int \exp \frac{i\, s(p,q)}{\lambda}\, dp, \qquad p \in \mathbb{C}^{n+k},\ q \in \mathbb{C}^n$$

is one of the main subjects in quantum physics [5,36,62]. It is known that the integral is asymptotic to 0 in order $\lambda^a\,(\log \lambda)^b$ (Nilsson class) as the wave length $\lambda$ tends to 0 at each $q$ [8]. It is known that when $\lambda$ is sufficiently small, the oscillation is very quick away from the critical point of $s(p,q)$ in $p$ and the behavior of $s(p,q)$ at a regular point $p$ does not contribute to the primary part of the asymptotic expansion as $\lambda$ tends to 0.

The stationary phase method [7,36,62] suggests that the principal term of the asymptotic expansion is determined up to mod $\pi/2$ (at least at a generic point $q$ where the function $s(p,q)$ is a Morse function) by the geometry of the family of the *wave front* $D_d \subset \mathbb{C}^n$ in the configuration space parameterized by $d \in \mathbb{C}$ defined as follows:

$q \in D_d$ if and only if $d$ is one of the critical values of the function $s(*,q)$.

The simplest example is

$$\int_{-\infty}^{\infty} \exp \pm \frac{ip^2/2}{\lambda}\, dp = \exp \pm \frac{\pi}{4} i \sqrt{2\pi\lambda}\ ,$$

which follows immediately from the formula

$$\int_{-\infty}^{\infty} \exp - \frac{p^2/2}{k}\, dp = \sqrt{2\pi k}$$

for Re $k > 0$, and by taking the limit as $k$ tends to $\mp i\lambda$. The signature of the potential function $S = \pm \frac{p^2}{2\lambda}$ affects the phase to give the difference $\pm \frac{\pi}{4}$ in phase (see [36]). The first non trivial example is the Airy-Fock's function $(Ai_1 = Ai)$

$$Ai_\lambda(q) = \int_{-\infty}^{\infty} \exp \frac{i(p^3/3 + pq)}{\lambda}\, dp,$$

which has the asymptotic expansion for real $q$.

$$Ai_\lambda(q) = \begin{cases} C_1 \dfrac{\lambda^{1/2}}{|q|^{1/4}} \sin\left(\dfrac{2}{3}\dfrac{-|q|^{3/2}}{\lambda} + \dfrac{\pi}{4}\right) + C_2\, \lambda^{3/2} + \cdots & (q < 0) \\ C_3 \lambda^{1/3} & (q = 0) \\ C_4 \dfrac{\lambda^{1/2}}{q^{1/4}} \dfrac{1}{\exp \frac{2}{3}\frac{q^{3/2}}{\lambda}} + C_5\, \lambda^{3/2} + \cdots & (q > 0) \end{cases}$$

as $\lambda \to 0$ [27]. Here $\pm\frac{2}{3}\frac{q^{3/2}}{\lambda}$ is the critical value of the potential function $\frac{(p^3/3+pq)}{\lambda}$ and the expansion in the case $q < 0$ is obtained by the sum of the asymptotic expansions at two critical points. The phase constant $\frac{\pi}{4}$ as above was already seen in the first example.

This integral is the unique solution of Airy's equation

$$I'' - \lambda^{-2/3}qI = 0.$$

By the relation $Ai_\lambda(q) = \lambda^{1/3}Ai(\frac{q}{\lambda^{2/3}})$ the asymptotics as $\lambda \to 0$ is concentrated in the local behavior of the solution at $\infty$, where the formal solution has Stokes phenomena to present a gap between positive and negative half lines. This causes the different asymptotic expansions on the positive and negative sides of the $q$-line.

The asymptotics of Pearcey's integral

$$I_\lambda(q,r) = \int_{-\infty}^{\infty} \exp\frac{i(p^4 + qp^2 + rp)}{\lambda}\,dp$$

was studied by Pearcey [60]. But the asymptotics is not yet well understood. The wave fronts associated to this integral is modeled on spherical optics, and the integral approximates the intensity of light nearby the cusp of the caustics. We call the family $W = \{D_d\}$ the *web* since the wave fronts form a configuration of foliations at generic points. In this note some differential geometric properties of webs are investigated and the various problems from the view point of web geometry are proposed.

We apply the singularity theory to our holonomic PDEs with first integrals and prove the existence of the (uni)versal model for almost all webs (Theorem 19.2).

Since the asymptotics of the oscillatory integral is determined by local behavior of $s(p,q)$ at critical points, we may allow the domain of integration to be a singular space parameterized by $q$. We study the generalized oscillatory integrals

$$I_\lambda(q) = \int_{V_q} \exp\frac{i\,s(p,q)}{\lambda}\,dp, \qquad q \in \mathbb{C}^n,$$

where $V_q$ is a variety parameterized by $q \in \mathbb{C}^n$ and $s(*,q)$ is a function on $V_q$. We show that all first order holonomic PDEs with first integrals are realized as webs of the generalized oscillatory integrals (Theorem 18.1). A result due to Goryunov [34] tells the complement of the singular locus $\mathrm{Sing}(W)$ (= Caustics + some other components, see sect. 17 for the definition) of the generalized integrals is an Eilenberg-Maclane's $K(\pi,1)$-space: the fundamental groups $\pi$

are finite index subgroups of the braid group $B(m)$ of $m$ strings. The number $m$ is the *Milnor number* of $s(*, q)$, i.e. the number of critical points of a generic perturbation of $s(*, 0)$, and the index is the local intersection number of $m$ solutions. We apply web geometry to define the affine connection on the configuration space $\mathbb{C}^n$ and show that in some cases the structure of the webs are determined by their curvature forms (Theorem 26.3). In analyzing the moduli of the normal forms of the webs, a certain residue of dynamics acting on the phase space $\mathbb{C}$ turns out to give a formal invariant. In sect. 24, 25 we discuss normal forms of holonomic PDEs with first integrals.

## 2  Radon transformation or Twistor theory

Here I will introduce the method to produce all holonomic PDEs with complete first integrals defined by smooth varieties $V$ in the first jet space. Consider a divergent diagram of smooth mappings

$$\mathbb{C}, 0 \xleftarrow{\lambda} \mathbb{C}^{n+k}, 0 \xrightarrow{f} \mathbb{C}^n, 0 .$$

For a generic $t \in \mathbb{C}$, the fiber $\lambda^{-1}(t)$ is smooth of codimension 1, on which $f$ restricts to a stable mapping in the sense of the singularity theory of mappings [30,47,48]. The discriminant $D(f \mid \lambda^{-1}(t))$ denoted $D_t$ is a subvariety of codimension one in $\mathbb{C}^n$ parameterized by $t$. Define a PDE $V \subset J^1(\mathbb{C}^{n-1}, \mathbb{C})$ by the Nash blow up of the family: $V$ is the closure of the set of those $(x, y, p)$ such that the contact element defined by $dy - \Sigma\, p_i dx_i = 0$ is tangent to $V$ at smooth points $(x, y) \in V$. Clearly those $D_t$ satisfy the holonomic PDE $V$ on $\mathbb{C}^n$, and $t$ is its first integral. In this manner all PDEs with first integrals are obtained. From the dual point of view, the above divergent diagram is seen as a deformation of the functions

$$\lambda_p : V_p = f^{-1}(p) \subset \mathbb{C}^{n+k} \to \mathbb{C}$$

with parameter $p \in \mathbb{C}^n$. Now consider the map

$$F = (f, \lambda) : \mathbb{C}^{n+k} \to \mathbb{C}^{n+1} .$$

This map is singular at a $Q$ if and only if either (1): one of $f, \lambda$ is singular at $Q$ or (2): $f, \lambda$ are non singular but their fibers are not transverse at $Q$. The critical point set of $f$ is, generically, of dimension $n-2$ and that of $F$ is of dimension $n-1$. So at a generic point of the singular locus $\Sigma(F)$, the later condition (2) holds, and it is seen that

$$D_t \ni p \text{ if and only if } t \text{ is a critical value of } \lambda_p.$$

Here the source $f^{-1}(p)$ is possibly singular and the value of $\lambda$ at those singular points are said to be critical values.

More generally it would be interesting to consider divergent diagrams of smooth mappings

$$\mathbb{C}^n, 0 \xleftarrow{\lambda} \mathbb{C}^p, 0 \xrightarrow{f} \mathbb{C}^q, 0,$$

$p$ being sufficiently large. In the case $n = q$, the discriminant sets $D_x \subset \mathbb{C}^q$ parameterized by $x \in \mathbb{C}^n$ and the duals $D_y \subset \mathbb{C}^n$ parameterized by $y \in \mathbb{C}^q$ satisfy second order partial differential equations respectively on $\mathbb{C}^q$ and $\mathbb{C}^n$.

## 3  Versal deformation of $\lambda_0$ and versal PDE

Thom-Mather theory applies to embed the family of functions $\lambda_p, p \in \mathbb{C}^n$ into a versal (roughly stating, sufficiently big) family $\lambda_{\tilde{p}}, \tilde{p} \in \mathbb{C}^{n+s}$, which is uniquely determined by the function $\lambda_0$ on $V_0$, $0 \in \mathbb{C}^n$. In the same manner as in the previous section, to the versal family $\{\lambda_{\tilde{p}}\}$ there corresponds a PDE $\tilde{V}$ defined on $\mathbb{C}^{n+s}$.

We define the versal PDE by the following commutative diagram.

PDE on $\mathbb{C}^n$ with 1st integral $\lambda$ $\longleftarrow$ Deformation $\lambda_x$

embedding $\downarrow$             embedding $\downarrow$

Versal PDE on $\mathbb{C}^{n+r}$ with 1st integral $\tilde{\lambda}$ $\longleftarrow$ Versal deformation $\lambda_{x,u}$.

By construction the structure of solutions of $V$ is seen on the transversal section of $\mathbb{C}^{n+s}$ by $\mathbb{C}^n$. The classification of our PDEs with first integrals is therefore equivalent to the classification of transverse sections of family of solutions of versal PDEs. More precisely the moduli space of a PDE with first integral is

*moduli of $V_0$ and the function $\lambda_0$ on $V_0$*

+ *moduli of (transverse) embeddings of $\mathbb{C}^n$ into $\mathbb{C}^{n+k}$ (mod some relation)* .

The first factor is of finite dimension (= Milnor number of $\lambda_0$) but the second factor is of infinite dimension and called *function moduli*.

In the case where the dimension $n$ of the configuration space is not large, at most $(n + 1)$-web structure appears in a family of solutions of a generic PDE with first integral on $\mathbb{C}^n$. This phenomenon is based on the nature of the function space, that is, the set of all modular (non simple) functions on varieties has sufficiently large codimension in the space of functions. This gives one reason to study $(n + 1)$-web structure of PDEs such as in sect. 14. Blaschke curvature form has various remarkable properties in this case. For example it is shown that in some cases the set of the Blaschke curvature forms is in one to one correspondence to the functional moduli (Theorem 26.3).

## 4  Bott connection and transverse dynamics

Let $\omega$ be a germ of non singular one form on a surface $M$. The integral manifolds of $\omega$ form a non singular foliation $F$ of codimension 1. By Frobenius theorem or the division theorem, there exists a one form $\theta$ such that

$$d\omega = \theta \wedge \omega$$

holds. Since

$$d\theta \wedge \omega = d\theta \wedge \omega - \theta \wedge (\theta \wedge \omega) = d\theta \wedge \omega - \theta \wedge d\omega = d(d\omega) = 0$$

$d\theta$ is closed on each leaf. Let $U \subset M$ be a connected open subset (flow box) such that the intersections of leaves with $U$ are simply connected. Then the integral on the leaves $\int \theta$ is well defined. Now define $\tilde{\omega} = \exp\left(-\int \theta\right)\omega$. Then $\tilde{\omega}$ is closed as

$$d\,\tilde{\omega} = d\left(\exp\left(-\int \theta\right)\omega\right) = d\exp\left(-\int \theta\right) \wedge \omega + \exp\left(-\int \theta\right) \wedge d\omega = 0.$$

Clearly $\tilde{\omega}$ defines the same foliation as $\omega$. The integral $f = \int \tilde{\omega}$ is well defined on $U$ and is a local defining level function of the foliation.

Now take a non singular function $x$ on $U$ such that $df \wedge dx \neq 0$, by shrinking $U$ if necessary. Then the mapping $(x, f)$ is a diffeomorphism of $U$ into $xy$-plane which sends the leaves to the horizontal lines defined by $dy = 0$ on the target space. We call such $U$ a flow box (this definition is slightly different from the standard one).

*Transverse dynamics* is a germ of a diffeomorphism from a transverse section $T$ of a leaf $L$ to another transverse section $T'$.

To define the dynamics, first assume that these intersection points are in the domain $U$ and on the line $y = 0$. Then the dynamics $S$ is defined by

$$S(x, y) = (x', y), \qquad (x, y) \in T, \quad (x', y) \in T' .$$

In general, the dynamics is defined by some composition of the above dynamics, by choosing a chain of transverse sections $T_1, \ldots, T_k$ of a leaf $L$ such that each pair $(T_i, T_{i+1})$ is contained in a flow box. Differential of the transverse dynamics in the normal direction tells how to translate a normal vector along leaves. This is the parallel translate of Bott connection.

Bott connection $\nabla$ is a (partial) connection of the normal bundle of the foliation defined by

$$\nabla_X Y = [X, Y],$$

where $X$ is a vector field along leaves and $Y$ is a section of the normal bundle. It is seen that $\nabla_X Y(p)$ is determined by $X(p)$ and the 1-jet of $Y$ at $p$. The parallel translate of normal vector space is defined only in the direction of the leaf. This is why this is called a partial connection. Normally this is extended with some additional data to an affine connection which preserves the tangent distribution of the leaves to define the various topological invariants such as Chern classes. In web geometry, such extensions were defined in a natural manner by Cartan, Chern, Blaschke and Goldberg. We will discuss this later on for 3-webs of codimension 1 on a surface in sect. 5 and $(n + 1)$-webs of codimension 1 on an $n$-space in sect. 14.

Here the author would emphasize the meaning of Bott connection by using the terminology of the dual frame of $\omega$ to enable us to understand the idea of Chern connection of 3 web defined in the next section. The one form $\omega$ defines a coframe $V_p \in T_p M/T_p F$. In terms of this coframe a normal vector $X \in T_p M/T_p F$ is presented as

$$X = \omega_p(X) \cdot V_p,$$

where $V_p$ is the dual frame of $\omega_p$: $\omega_p(V_p) = 1$. To find the translate $Y = S(X)$ of $X$ along a leaf $L$ from $p$ to $q$, it is enough to obtain the coefficient $\omega_q(Y)$.

This can be calculated as follows. First assume that $p, q$ are in a flow box, and denote $\omega = f \, dy$ in the coordinate as above. Then the dual frame is $1/f \, \partial/\partial y$. Clearly $Y$ has the same proportion as $X$ in terms of $y$ by definition. Therefore

$$\omega_q(Y) = \frac{f(q)}{f(p)} \, \omega_p(X) \, .$$

On the other hand $d\omega = \theta \wedge \omega = d \log f \wedge \omega$. Therefore

$$\frac{f(q)}{f(p)} = \exp \int_p^q \theta,$$

where the path of integration is unique up to homotopy as the leaf in flow box is simply connected. In general we obtain the formula

$$\omega_q(S(X)) = \exp \left( \int_p^q \theta \right) \omega_p(X),$$

where $\int_p^q$ stands for the integration along a path on a leaf $L$ from $p$ to $q$.

## 5    Chern connection of 3-webs of dimension 1

Now assume a germ of a 3-web $W$ on a surface $M$ is defined by a triple of germs of 1-forms $\omega_1, \omega_2, \omega_3$ mutually linearly independent. Since 3 covectors in 2-space has a linear relation, by multiplying these defining one forms by unit functions, we may assume the normalization condition

$$\omega_1 + \omega_2 + \omega_3 = 0,$$

and then it follows

$$\omega_1 \wedge \omega_2 = \omega_2 \wedge \omega_3 = \omega_3 \wedge \omega_1.$$

Now by division of $d\omega_i$ by $\omega_i$ we obtain one form $\theta_i$ such that

$$d\omega_i = \theta_i \wedge \omega_i.$$

Since $\theta_i$ has ambiguity of type $f_i\omega_i$, we may find $f_1, f_2$ such that

$$\theta_1 + f_1\omega_1 = \theta_2 + f_2\omega_2$$

holds. Denote this 1-form by $\theta$. Then

$$d\omega_i = \theta \wedge \omega_i$$

holds for $i = 1, 2, 3$. These equations may be presented as

$$d \begin{pmatrix} \omega_1 \\ \omega_2 \\ \omega_3 \end{pmatrix} = \begin{pmatrix} \theta & 0 & 0 \\ 0 & \theta & 0 \\ 0 & 0 & \theta \end{pmatrix} \wedge \begin{pmatrix} \omega_1 \\ \omega_2 \\ \omega_3 \end{pmatrix} \qquad (*)$$

Each 2-lines (say $j, k$-lines, $j, k \neq i$) of this equation defines an affine connection $\nabla_i$ without torsion by the structure equation

$$d \begin{pmatrix} \omega_j \\ \omega_k \end{pmatrix} = \begin{pmatrix} \theta & 0 \\ 0 & \theta \end{pmatrix} \wedge \begin{pmatrix} \omega_j \\ \omega_k \end{pmatrix}$$

with coframe $(\omega_j, \omega_k)$ and connection form $\begin{pmatrix} \theta & 0 \\ 0 & \theta \end{pmatrix}$. By the form of structure equation, it can be seen that this connection extends Bott connections of $\omega_j$ and $\omega_k$ simultaneously.

Geometrically the parallel translation $T$ of tangent spaces of the surface along a leaf is seen as the translate of the infinitesimally small triangles $\Delta$ to $\Delta'$, where horizontal and vertical lines are leaves of $\omega_j, \omega_k$ and the transverse lines are leaves of $\omega_i$.

The above equation $(*)$ shows the affine connection is independent of $i$. We call this affine connection $\nabla_i$ *Chern connection*.

We proved

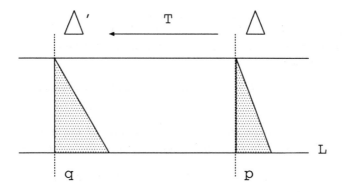

Figure 1.

**Theorem 5.1** *Chern connection of a non singular 3-web $W = (F_1, F_2, F_3)$ is the unique common extension of Bott connection of $F_i$ to an affine connection.*

The connection form $\theta$ depends on the defining one forms $\omega_i$, but the curvature form

$$\Omega = d \begin{pmatrix} \theta & 0 \\ 0 & \theta \end{pmatrix} = \begin{pmatrix} d\theta & 0 \\ 0 & d\theta \end{pmatrix}$$

is independent of the choice of $\omega_i$. Classically $d\theta$ (instead of the matrix) is called the *web curvature form* in the web geometry.

The web curvature form has a significant property as follows.

It is classically known that if the curvature form vanishes identically, $W$ is diffeomorphic to the (hexagonal) web defined by parallel lines with three distinct directions.

And also quite recently the next theorem was proved by the author.

**Theorem 5.2** ([59]) *Let $W = (F_1, F_2, \ldots, F_d)$ be a d-web on a neighbourhood of the origin in $\mathbb{C}^2$ defined by meromorphic 1-forms. Assume 3-subwebs of $W$ are non flat, i.e. the curvature forms do not vanish identically. Then the following conditions are equivalent.*

*(1) $F_i$ is defined by a 1-form $\omega^{t_i}, t_i \in \mathbb{P}^1$, in a linear family of meromorphic 1-forms $P = \{\omega^t\}$ of dimension 1.*

162

(2) *The modulus of tangents to the leaves of $F_i, i = 1, ..., d$ passing through a point is constant on the complement of $\Sigma(W)$ where $\Sigma(W)$ is the set of those points where $W$ is not a configuration of $d$ non singular foliations in general position.*

(3) *$F_1, ..., F_d$ are weakly associative: 3-subwebs have equal web curvature form.*

(4) *$F_1, ..., F_d$ are associative: Bott connections of the normal bundles of $F_1, ..., F_d$ extend to equal affine connection on the complement of $\Sigma(W)$.*

The flat case is classically known by Poincaré, Reidemeister and Mayrhofer as follows [50,51,63].

**Theorem 5.3** *Let $W = (F_1, ..., F_d)$ be a germ of a d-web on $\mathbb{C}^2$. Assume Chern connections of all 3-subwebs are flat and $d \neq 5$. Then $W$ is holomorphically diffeomorphic to a germ of a d-web of d pencils of lines.*

In the case $d = 5$ there is Bol's exceptional 5-web which is hexagonal and not linearizable but admits an Abelian relation similar to the 5-term relation of cross ratio [10].

## 6 Web curvature form of 3-webs

Assume that a germ of a 3-web on the complex plane at the origin is given by three 1-forms $dx, dy$ and $df$, $f$ being a germ of a holomorphic function. Replace these defining forms by $\omega_1 = -f_x dx, \omega_2 = -f_y dy$ and $\omega_3 = df$. Then

$$\omega_1 + \omega_2 + \omega_3 = 0$$

holds. The connection form for this normal form is

$$\theta = (\log f_x)_y dy + (\log f_y)_x dx$$

and the web curvature form is

$$d\theta = \frac{\partial^2}{\partial x \partial y} \log \frac{f_x}{f_y} \, dx \wedge dy .$$

Instead of $df$, take a germ of a holomorphic 1-form on $\mathbb{C}^2$

$$\omega_1 = adx + bdy$$

and let $\omega_2 = dx$, $\omega_3 = dy$, $x, y$ being the coordinate functions. The *polar curve $P$* of $\omega_1$ with respect to the projection onto the $x$-coordinate is the set of those points where the leaf of $\omega_1 = 0$ is tangent to that of $dx = 0$, in other words,

$$P = \{\omega_1 \wedge dx = -b \; dx = 0\}.$$

By Bertini-Sard theorem, for a generic coordinate function $x$, the polar curve is irreducible and reduced, hence the leaf of $\omega_1 = 0$ has contact of order 2 with $dx = 0$ at generic points of $P$. Also for the second coordinate $y$ we may assume that the polar curve

$$Q = \{\omega_1 \wedge dy = a \; dx \wedge dy = 0\}$$

is irreducible and reduced. Now normalize $\omega_1, \omega_2, \omega_3$ as follows:

$$\tilde{\omega}_1 = \omega_1, \quad \tilde{\omega}_2 = -a \; dx, \quad \tilde{\omega}_3 = -b \; dy.$$

The connection form of the 3-web of $\omega_1, \omega_2, \omega_3$ is

$$\theta = \frac{b_x}{b} dx + \frac{a_y}{a} dy$$

and

$$d\omega_i = \theta \wedge \omega_i, \qquad i = 1, 2, 3,$$

holds. By the above form of $\theta$, the connection form has a simple pole along $P, Q$, and also the web curvature form

$$\Omega = d\theta = (\log \frac{a}{b})_{xy} \; dx \wedge dy$$

has a pole of order 2 along the polar curves.

**Proposition 6.1** *Assume that the 3-web given by $\omega_1, dx, dy$ is flat: $d\theta = 0$. Then $\omega_1$ admits a first integral.*

*Proof.* By assumption, $a/b = A(x)/B(y)$, $A, B$ being meromorphic functions of one variable. So $\omega_1$ may be replaced with $A(x)dx + B(y)dy$. By integrating this form, we obtain a first integral. $\qquad\square$

## 7   Normal form of non singular plane 3-webs

Assume a germ of a non singular 3-web $W$ on the complex plane at the origin is given by three 1-forms $dx, dy$ and $df$, $f$ being a germ of holomorphic function. Now by coordinate change of $x, y$ we may assume

$$f(t, 0) = f(t.0) = t.$$

Next consider the dynamics acting on the $x$-coordinate line

$$(t,0) \to (t,t) \to (f(t,t),0).$$

It is easily seen that $f(t,t)$ has linear term $2t$. By Poincaré's theorem, this dynamics can be linearized by a coordinate change. This tells that by a coordinate change of the form $\phi \times \phi$ preserving the foliations by $dx, dy$, the dynamics geometrically defined above becomes linear: $f(t,t) = 2t$. By this normalization, we may assume

$$f(x,y) = x + y + k\ xy(x - y) + \text{h.o.t.} ,$$

where h.o.t. stands for a heigher order term of order $\geq 4$ in $x, y$ vanishing on the $x$ and $y$ axies. The web curvature form of the normal form at the origin is

$$d\theta(0,0) = 4kdx \wedge dy .$$

This normal form of $f$ admits a similar coordinate change $(x, y) \to (\lambda x, \lambda y)$, which normalizes the coefficient $4k$ to 1 if $k \neq 0$. At each point such a coordinate is uniquely determined and the 2-form $dx \wedge dy$ in the coordinate at each point defines a holomorphic 2-form on the surface, which is nothing but the web curvature form $d\theta$ of the web.

## 8  Geometric interpretation of the curvature form

Assume a germ of a 3-web on the complex plane at the origin is given by three 1-forms $dx, dy$ and $df$, $f$ being a germ of the normal form.

Consider the anti-clockwise move on the nest starting at $P = (x,0)$ as in the figure 2.

We call this diagram a Brianchon hexagon. By direct calculation

$$Q = (0, x)$$
$$R = (-x - 2kx^3, x)$$
$$S = (-x - 2kx^3, 0)$$
$$T = (0, -x - 2kx^3)$$
$$U = (x + 4kx^3, -x - 2kx^3)$$
$$V = (x + 4kx^3, 0) ,$$

modulo higher order terms of oeder $\geq 4$ in $x$. Therefore the return mapping $R$ has the Taylor expansion

$$R(x) = x + 4kx^3 + \text{higher order term}$$

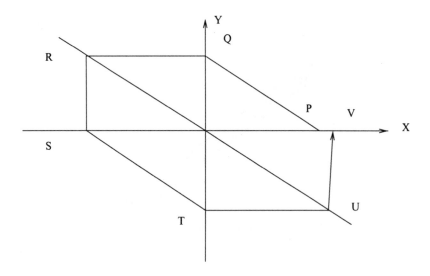

Figure 2.

To see this Taylor expansion without calculation, let us consider the parallel translate of the tangent plane at $P$ to $Q$ by Chern connection along the anti clockwise hexagon. Although the hexagon is not closed, we obtain

$$\int_{\text{Brianchon hexagon}} \theta = \int_{\text{"interior" of Brianchon hexagon}} d\theta + \text{higher order term in } x$$

The integral of $\theta$ gives the logarithm of the linear holonomy in the dual frames of $\omega$ at $V$ and $P$. Assume $\omega_1 = dx$. Then the linear holonomy is nothing but the derivative of the return map $R$ at $x$. Therefore

$$= \log R'(x)$$

Therefore

$$R'(x) = 1 + 4k \text{ Area of the hexagon } + \text{ higher order term}$$
$$= 1 + 12k \, x^2 + \text{ higher order term,}$$

hence

$$R(x) = x + 4k \, x^3 + \text{higher order term.}$$

Now consider the diagram in the figure 3. In the diagram an (infinitesi-

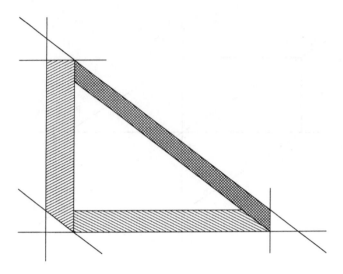

Figure 3.

mally thin) non closed Möbius band is seen. This is closed if and only if the connection is flat, i.e. the curvature form vanishes identitically. When the connection is flat, we may compare this diagram with the closed hexagonal diagram by shrinking the triangle inside the Möbius band to a point. In fact a 3-web is flat if and only if all Brianchon hexagons are closed.

## 9   Singular 3-webs on the plane

A germ of a singular 3-web on $\mathbb{C}^2$ at the origin is defined by a germ of a variety $V$ in the 1-jet space $J^1(\mathbb{C}, \mathbb{C})$ at the fiber of the origin such that the natural projection $\pi$ of $V$ onto $\mathbb{C} \times \mathbb{C}$ is generically 3-to-1. The *singular locus* $\Sigma(W)$ is the set of those $p \in \mathbb{C} \times \mathbb{C}$ such that $V$ is singular at the fiber of $p$ or non transverse to the contact elements. By the transversality the contact form $\omega = dy - pdx$ on the 1-jet space induces non singular one forms on $V$ at the fiber of $p \notin \Sigma(W)$, which project to the base $\mathbb{C} \times \mathbb{C}$ to form non singular foliations at $p$. Since $V$ is non singular at the fiber of $p$, those foliations are in general position. It is not difficult to see that there is a meromorphic function $f$ on $V$ such that the direct image of $f\omega$ vanishes identically on a neighbourhood of the origin. Denote the branches of $f\omega$ by $\omega_1, \omega_2, \omega_3$. Then the normalization condition

$$\omega_1 + \omega_2 + \omega_3 = 0$$

holds, and

$$\omega_1 \wedge \omega_2 = \omega_2 \wedge \omega_3 = \omega_3 \wedge \omega_1.$$

Therefore this form is well defined on the base space $\mathbb{C} \times \mathbb{C}$. When the projection of $V$ is proper, $f$ may be chosen so that it vanishes only at the preimage of the branched curve. For example, present the variety as

$$p = P_1(x,y), \quad P_2(x,y), \quad P_3(x,y),$$

and define

$$f = (P_1 - P_2)(P_2 - P_3)(P_3 - P_1)(P_{i+1} - P_{i+2}),$$

on the $i$-th sheet with a cyclic index of order 3. Clearly this is bounded and independent of the ordering of the branches. Then the *singular point set* $\Sigma(W)$ is the zero set of the 2-form.

The singularity of the connection form is supported on this set, and holonomy of all leaves of $\omega_i$ are determined by how they meet the singular locus.

## 10    Example: Envelope of cusps and tangent lines

Let a singular 3-web $W = (F_1, F_2, F_3)$ be defined by the following 1-forms:

$$\omega_1 = dy + \sqrt{x}dx,$$
$$\omega_2 = dy - \sqrt{x}dx,$$
$$\omega_3 = dy + f(x,y)dx, \quad f(0,0) \neq 0$$

By integrating, it is seen that $\omega_1, \omega_2$ define the branches of the 2-web of cusps $9(y-c)^2 = 4x^3, c \in \mathbb{C}$, which is singular along the $y$-axis (cf. figure 4).

Dufour [24] proved that the "ordinary" envelope of a one-parameter family of cusps is equivalent to the normal form. To see this in our theory of sect. 19, it suffices to notice that the 2-web corresponds to the unfolding of the singularity $\lambda = u^3$ on the line. The Jacobi's ideal

$$J(\lambda) = \mathcal{O}_u / < 3u^2 >$$

is generated by 1 and $u$. Therefore the versal unfolding in the sense of sect. 19 is

$$\tilde{\lambda} = u^3 + xu = y$$
$$f(u,x,y) = (x,y).$$

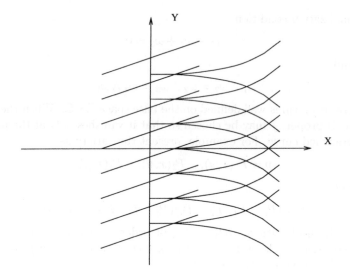

Figure 4.

It is seen that the PDE with first integral $\tilde{\lambda}$ corresponding to the versal unfolding is equivalent to the above normal form.

Now normalize these forms as follows (here we use a normalization different from the previous section):

$$\tilde{\omega}_1 = (\sqrt{x} + f)\omega_1, \quad \tilde{\omega}_2 = (\sqrt{x} - f)\omega_2, \quad \tilde{\omega}_3 = (-2\sqrt{x})\omega_3.$$

The connection form $\theta$ is

$$\theta = d\log\,(\sqrt{x} + f) + A\,\omega_1 = d\log\,(\sqrt{x} - f) + B\,\omega_2,$$

where

$$A = \frac{1}{f^2 - x}(\sqrt{x}f_y - \frac{f}{2x} + f_x), \qquad B = \frac{1}{f^2 - x}(\sqrt{x}f_y + \frac{f}{2x} - f_x).$$

The web curvature form is

$$d\theta = dA \wedge \omega_1 = dB \wedge \omega_2$$

$$= \frac{1}{(x - f^2)^2}\{\frac{f^3}{2}\frac{1}{x^2} + (\frac{f^2 f_x}{2} - f)\frac{1}{x}$$

$$+ \frac{3}{2}f_x + 2(xf_y^{2-f_x^2})f + (f^2 - x)(f_{xx} - xf_{yy})\}dx \wedge dy\}.$$

This has the pole of order 2 along $x = 0$ (and $x = f^2$) and the principal term along $x = 0$ is

$$\frac{1}{2f(0,y)x^2} \, dx \wedge dy.$$

## 11   Example: Two tangent foliations and a transverse foliation

Let a singular 3-web $W = (F_1, F_2, F_3)$ be defined by the following 1-forms:

$$\omega_1 = dy + x \, dx,$$
$$\omega_2 = dy,$$
$$\omega_3 = dy + f(x,y)dx, \quad f(0,0) \neq 0$$

Clearly $\omega_1, \omega_2$ define non singular foliations, which have contact of order 2 along the $y$-axis:

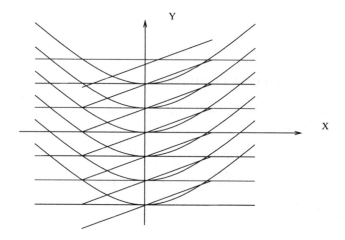

Figure 5.

Now normalize these forms as follows:

$$\tilde{\omega}_1 = f\omega_1, \quad \tilde{\omega}_2 = (x - f)\omega_2, \quad \tilde{\omega}_3 = -x\omega_3.$$

The connection form $\theta$ is

$$\theta = d \log f + A \, \omega_1$$
$$= d \log (x - f) + B \, \omega_2,$$

where

$$A = \frac{f - x f_x}{f_x(x - f)}, \qquad B = A + \frac{f_y}{f} + \frac{f_y}{x - f}.$$

The connection form $\theta$ has pole along the singular locus $x = 0$ and has the principal term

$$-\frac{1}{f(0,y)x} \, dy.$$

The curvature form is

$$d\theta = dB \wedge dy = B_x \wedge dy,$$

which has the pole of order 2 along $x = 0$, $x = f$ and the principal term is

$$\frac{1}{f(x,0)x^2} \, dx \wedge dy.$$

## 12   Example: 3 tangent foliations

Assume that $f = xg(x,y)$ and $g(0,0) \neq 0, 1$ in the above example. Then the connection form is

$$\theta = d\log (x - f) + B dy$$
$$= \frac{dx}{x} + d\log (1 - g) + B dy.$$

The principal term of the connection form is

$$\frac{dx}{x} - \frac{g_x(0,y)}{g(1 - g)(0,y) \, x} dy$$

and the principal term of the curvature form is

$$\frac{g_x(0,y)}{g(1 - g)(0,y) \, x^2} dy.$$

## 13 Example: Envelope of smooth curves and a transversal foliation

Let a singular 3-web $W = (F_1, F_2, F_3)$ be defined by the following 1-forms:

$$\omega_1 = dy + \frac{1}{\sqrt{x}}\, dx,$$

$$\omega_2 = dy - \frac{1}{\sqrt{x}}\, dx,$$

$$\omega_3 = dy + f(x, y)dx.$$

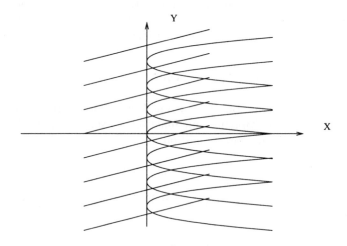

Figure 6.

To see this is a normal form of the envelope of smooth curves, and of a transversal foliation, it suffices to notice that the 2-web corresponds to the unfolding of the function $\lambda = u$ on the 2-line

$$V = \{(u + v)(u - v) = 0\}.$$

Let $f = (u+v)(u-v) : \mathbb{C}^2 \to \mathbb{C}$. Then the Jacobi ideal $J(f, \lambda)$ of the divergent diagram $(f, \lambda) : \mathbb{C} \xleftarrow{\lambda} \mathbb{C}^2 \xrightarrow{f} \mathbb{C}$ is generated by

$$(1, 0) \in \theta(\lambda) \oplus \theta(f) = \mathcal{O}_{u,v} \oplus \mathcal{O}_{u,v}$$

Therefore the versal unfolding with one parameter $y$ in the sense of sect. 19 is

$$\tilde{\lambda} = u + y$$
$$\tilde{f}(u,v,y) = ((u+v)(u-v),y).$$

It is seen that the PDE on the target space of $\tilde{f}$ with first integral $\tilde{\lambda}$ is equivalent to the above normal form.

Then the connection form is $d\theta$ and extends holomorphically to the singular locus $x = 0$. The extension vanishes identically on the singular locus. To see this one may use another coordinate system such that $\omega_3 = dy$ (of course $\omega_1, \omega_2$ are then different from the above form.) The lift of the 3-web by the map $(x^2, y)$ is non singular and symmetric. By the symmetry, the curvature form vanishes at the fixed point set $x = 0$. There the curvature form of the quotient vanishes also along the branched locus.

## 14   Chern connection of $(n+1)$-webs

Let $F_i$ be a germ of foliation of $\mathbb{C}^n$ defined by a holomorphic 1-form $\omega_i$ for $i = 1, \ldots n+1$. By multiplying unit functions to those defining 1-forms, we may assume the normalization condition

$$\omega_1 + \cdots + \omega_{n+1} = 0 .$$

By the integrability of those forms and Frobenius theorem, there exists a 1-forms $\theta_i$ such that

$$d\omega_i = \theta_i \wedge \omega_i$$

for $i = 1, \cdots, n+1$. These equations are presented as the structure equation of an affine connection with coframe $(\omega_1, \ldots, \hat{\omega}_i, \ldots, \omega_{n+1})$ on $\mathbb{C}^n$

$$d \begin{pmatrix} \omega_1 \\ \omega_2 \\ \cdots \\ \omega_{n+1} \end{pmatrix} = \begin{pmatrix} \theta_1 & 0 & \cdots & 0 \\ 0 & \theta_2 & \cdots & 0 \\ & & \cdots & \\ 0 & \cdots & 0 & \theta_{n+1} \end{pmatrix} \wedge \begin{pmatrix} \omega_1 \\ \omega_2 \\ \cdots \\ \omega_{n+1} \end{pmatrix} \qquad (*)$$

$$= \begin{pmatrix} \theta_i & 0 & \cdots & 0 \\ 0 & \theta_i & \cdots & 0 \\ & & \cdots & \\ 0 & \cdots & 0 & \theta_i \end{pmatrix} \wedge \begin{pmatrix} \omega_1 \\ \omega_2 \\ \cdots \\ \omega_{n+1} \end{pmatrix} + T$$

with torsion part

$$T = \begin{pmatrix} (\omega_1 - \omega_i) \wedge \omega_1 \\ (\omega_2 - \omega_i) \wedge \omega_2 \\ \cdots \\ (\omega_{n+1} - \omega_i) \wedge \omega_{n+1} \end{pmatrix}.$$

(Compare this with $(*)$ in sect. 5.) Of course $\theta_i$ has ambiguity of the form $g\omega_i$ and the connection is not uniquely defined. In order to kill the ambiguity, we assume the following normalization condition.

Write the torsion part as

$$T = \begin{pmatrix} T_{11}\omega_1 \wedge \omega_1 + T_{12}\omega_2 \wedge \omega_1 + \cdots + T_{1n}\omega_{n+1} \wedge \omega_1 \\ T_{21}\omega_1 \wedge \omega_2 + T_{22}\omega_2 \wedge \omega_2 + \cdots + T_{2,n+1}\omega_{n+1} \wedge \omega_2 \\ \cdots \\ T_{n+1,1}\omega_1 \wedge \omega_{n+1} + T_{n+1,2}\omega_2 \wedge \omega_{n+1} + \cdots + T_{n+1,n+1}\omega_{n+1} \wedge \omega_{n+1} \end{pmatrix},$$

where the indices $j, k$ of $T_{jk}$ avoid $i$ and $T_{j,j} = 0$ for $j = 1, \ldots, n + 1$. If we replace $\theta_i$ by a connection form

$$\theta_i + g\omega_i = \theta_i - g(\sum_{j \neq i} \omega_j),$$

then the torsion $T_{j,k}$ becomes

$$T_{jk} + g \qquad \text{for } j \neq k.$$

By the normalization condition

$$\sum_{i \neq j} T_{j,k} + g = 0 \qquad\qquad (T)$$

$g$ and the connection form $\theta_i$ are uniquely determined.

The $i$-th affine connection $\nabla_i$ is defined by the structure equation $(*)$ with the connection form $\theta_i I$, $I$ being the identity. By the structure equation $(*)$, we obtain

**Theorem 14.1** *For a non singular $(n + 1)$-web $W = (F_1, \ldots, F_{n+1})$ on $\mathbb{C}^n$, Bott connection of each foliation $F_i$ extends to unique affine connection $\nabla_i$.*

**Theorem 14.2** *The mean of curvature forms of the affine connections $\nabla_i$ is a constant multiple of Blaschke curvature form $d\Gamma$ defined in the next section.*

*Proof.* We use the notations of the previous section. Let $\omega_i = f_i du_i$ for $i = 1, \ldots, n+1$. Clearly the normalization condition

$$\omega_1 + \omega_2 + \cdots + \omega_{n+1} = 1$$

holds. We calculate the connection form $\theta_{n+1}$ of the $n+1$-st affine connection in terms of the web function $f$. We begin with the equation

$$\begin{aligned}
d\omega_i &= d(\log f_i) \wedge \omega_i \\
&= (d(\log f_{n+1}) + g\omega_{n+1}) \wedge \omega_i + T_i \wedge \omega_i .
\end{aligned}$$

Here

$$\begin{aligned}
T_i &= d(\log f_i - \log f_{n+1}) - g\omega_{n+1} \\
&= \frac{df_i}{f_i} - \frac{df_{n+1}}{f_{n+1}} - g\omega_{n+1} \\
&= \sum_{j=1,\ldots,n+1} \frac{f_{ij}}{f_i} du_j - \sum_{j=1,\ldots,n+1} \frac{f_{n+1,j}}{f_{n+1}} du_j - g\omega_{n+1}
\end{aligned}$$

$$= \sum_{j=1,\ldots,n+1} \frac{f_{ij}}{f_i f_j} \omega_j - \sum_{j=1,\ldots,n+1} \frac{f_{n+1,j}}{f_{n+1} f_j} \omega_j - g\omega_{n+1}$$

Since $\omega_1 + \omega_2 + \cdots + \omega_{n+1} = 1$,

$$= \sum_{j=1,\ldots,n+1} \left\{ \frac{f_{ij}}{f_i f_j} - \frac{f_{n+1,j}}{f_{n+1} f_j} - \frac{f_{i,n+1}}{f_i f_{n+1}} + \frac{f_{n+1,n+1}}{f_{n+1} f_{n+1}} + g \right\} \omega_j.$$

Therefore the torsion coefficient is

$$T_{ij} = \frac{f_{ij}}{f_i f_j} - \frac{f_{n+1,j}}{f_{n+1} f_j} - \frac{f_{i,n+1}}{f_i f_{n+1}} + \frac{f_{n+1,n+1}}{f_{n+1} f_{n+1}} + g$$

for $i, j = 1, \ldots, n$ and $i \neq j$, from which

$$\begin{aligned}
\sum_{i,j=1,\ldots,n, i \neq j} T_{ij} = &\sum_{i,j=1,\ldots,n+1} \frac{f_{ij}}{f_i f_j} - \sum_{i=1,\ldots,n+1} \frac{f_{i,i}}{f_i f_i} \\
&- 2n \sum_{i=1,\ldots,n+1} \frac{f_{n+1,i}}{f_{n+1} f_i} + (n^2 + n) \frac{f_{n+1,n+1}}{f_{n+1} f_{n+1}} + (n^2 - n)g.
\end{aligned}$$

Now the normalization condition (T) implies

$$g = \frac{1}{n-n^2}\left\{ \sum_{i,j=1,\ldots,n+1} \frac{f_{ij}}{f_i f_j} - \sum_{i=1,\ldots,n+1} \frac{f_{i,i}}{f_i f_i} \right.$$

$$\left. -2n \sum_{i=1,\ldots,n+1} \frac{f_{n+1,i}}{f_{n+1} f_i} + (n^2+n)\frac{f_{n+1,n+1}}{f_{n+1} f_{n+1}} \right\}.$$

Therefore the connection form of $\nabla_{n+1}$ is

$$d(\log f_{n+1}) + g\omega_{n_1} = d(\log f_{n+1})$$

$$+ \frac{1}{n-n^2}\left\{ \sum_{i,j=1,\ldots,n+1} \frac{f_{ij}}{f_i f_j} - \sum_{i=1,\ldots,n+1} \frac{f_{i,i}}{f_i f_i} \right\}\omega_{n+1}$$

$$+ \frac{2}{n-1} \sum_{i=1,\ldots,n+1} \frac{f_{i,n+1}}{f_i}du_{n+1} - \frac{n+1}{n-1}\frac{f_{n+1,n+1}}{f_{n+1} f_{n+1}}du_{n+1}.$$

The mean connection form of $\nabla_j$ for $j=1,\ldots,n+1$ is

$$\frac{1}{n+1}\left\{ d(\log f_1 \cdots f_{n+1}) + \frac{2}{n-1} \sum_{i,j=1,\ldots,n+1} \frac{f_{ij}}{f_i}du_j \right.$$

$$\left. -\frac{n+1}{n-1} \sum_{1\leq j\leq n+1} (\log f_j)_j du_j \right\}$$

$$= \frac{1}{n+1}\left\{ (1+\frac{2}{n-1})d(\log f_1 \cdots f_{n+1}) - \frac{n+1}{n-1} \sum_{1\leq j\leq n+1} (\log f_j)_j du_j \right\}$$

hence the mean curvature form is

$$-\frac{1}{n-1} \sum_{i,j=1,\ldots,n+1} (\log f_j)_{ij} du_i \wedge du_j$$

$$= \frac{1}{n-1}\cdot\frac{1}{2} \sum_{i,j=1,\ldots,n+1} \log \frac{f_i}{f_j})_{ij} du_i \wedge du_j = \frac{1}{n-1}d\Gamma.$$

## 15 Chern connection and Blaschke curvature form

Before differential geometry of webs was systematically investigated by Russian school, Blaschke introduced a curvature form for non singular $(n+1)$-webs on $n$-space $\mathbb{C}^n$. Assume $F_i$ is defined by a defining level function $u_i$. Since those defining functions are non singular and $du_i$ are in general position, the map $(u_1, \ldots, u_{n+1})$ is a germ of an immersion of $\mathbb{C}^n$ into $\mathbb{C}^{n+1}$. Let $f = 0$ be a defining equation of the image. Such an $f$ is called a *web function*. Here

$$f(u_1, \ldots, u_{n+1}) = 0$$

holds identically. *Blaschke curvature form* is then defined by

$$d\Gamma = \frac{1}{2} \sum_{i,j} \frac{\partial^2}{\partial u_i \partial u_j} (\log \frac{f_i}{f_j}) \, du_i \wedge du_j,$$

where $f_i$ stands for the partial derivative of $f$ by $u_i$. It is seen that the curvature form is independent of the choice of level functions and the web function. This curvature form is $n - 1$ times the curvature form of the mean connection of Chern connections $\nabla_1, \ldots, \nabla_{n+1}$ defined in sect. 14. So we define the Blaschke connection by the $n - 1$ time the mean connection of Chern connections $\nabla_1, \ldots, \nabla_{n+1}$.

**Theorem 15.1** *For an $(n + 1)$-web, the curvature form of Blaschke connection is diagonal with equal entries, and the entry is Blaschke curvature form.*

In terms of Blaschke curvature form Theorem 5.2 is generalized as follows.

**Theorem 15.2** ([59]) *Assume $d$ foliations $F_1, \ldots, F_d$, $n + 2 \leq d$ of codimension 1 on an $n$-manifold are non singular, in general position and also the curvilinear 3-webs on the intersection of leaves of $n - 2$ foliations cut out by 3 of the remaining 4 foliations are non hexagonal. Then the following conditions are equivalent.*

(1) $F_1, \ldots, F_d$ *are associative: the modulus of tangent planes to the leaves of $F_1, \ldots, F_d$ passing through a point is constant.*

(2) $F_1, \ldots, F_d$ *are weakly associative: Blaschke curvature forms of $(n + 1)$-subwebs are equal.*

## 16    Chern connection and Blaschke curvature form

First we will define the curvature form of $d$-webs on an $n$-space for all $m \geq n+1$. The idea is as follows. By Theorem 14.2, Blaschke curvature form of an $(n + 1)$-form is the mean curvature of Chern connections $\nabla_i, i = 1, \ldots, n + 1$, up to scalar multiplication.

Now let $W = (F_1, \ldots, F_d)$ be a $d$-web of codimension 1 and $d \geq n+1$. We define the *Blaschke connection form of $W$* by the mean of Blaschke connection forms of all $(n+1)$-(non marked) subwebs. We define the $i$-th connection $\nabla_i$ by the mean of Chern connections of all marked $(n+1)$-subwebs $(F_{s_1}, \ldots, F_{s_{n+1}})$ such that one of $F_{s_j}$ is $F_i$ and specialized.

In the case $d \geq n+2$ the connection form of $\nabla_i$ is not the mean of the connection forms of all $(n+1)$-subwebs with $F_i$ as a specialized member. And also the connection form of Blaschke connection is not the mean of the connection forms of all non marked $(n+1)$-subwebs. This is the case if the modulus of the tangent planes of the leaves of the foliations passing through a point is constant. But if it is the case, all foliations $F_1, \ldots F_d$ embed to a linear family of one forms. If all members of the linear family are integrable, Blaschke curvature of $(n+1)$-subwebs vanishes identitically.

By definition we obtain

**Proposition 16.1** *Let $W = (F_1, \ldots, F_d)$ be a germ of a $d$-web on $\mathbb{C}^n$, $n+1 \leq d$. The affine connection $\nabla_i$ extends Bott connection of $F_i$ for $i = 1, \ldots, d$. The mean of the connections $\nabla_i$ is $\frac{1}{n-1}$ Blaschke connection.*

Let $V$ be an analytic subvariety of the 1-jet space $J^1(\mathbb{C}^{n-1}, \mathbb{C}^1)$ of dimension $n$, on which the natural projection $\pi$ of the jet space onto the base space $\mathbb{C}^{n-1} \times \mathbb{C}^1$ restricts to a proper and $d$-to-1 mapping. In this section we assume the restriction $\omega|_V$ of the contact form $\omega = dy - \sum_{i=1,\ldots,n-1} p_i dx_i$ to $V$ is locally integrable on the complement of the singular locus of $V$ and that of the restriction.

**Theorem 16.1** *The Bott connection of the foliation on the non-singular locus of $\omega|_V$ extends to a unique affine connection $\nabla$ of the smooth part. The connection form $\Theta$ is diagonal with equal entry $\theta$. The curvature form $d\Theta$ extends also to a diagonal meromorphic 2-form on $V$. The direct image of $\Theta$ is the connection above defined for the $d$-web structure on the base space $\mathbb{C}^{n-1} \times \mathbb{C}$. The singularity of the direct image is supported by the singularity of the web, where the web is not non singular and in general position.*

And in the case $d = n+1$, we obtain

**Proposition 16.2** *Assume $d = n+1$. Then there exists an analytic function $f$ on $V$ such that the direct image of $f\omega$ vanishes identically. Such a function $f$ is unique up to a multiple by pull back of functions on $\mathbb{C}^{n-1} \times \mathbb{C}$, and defines a class in the space of finite dimension $\mathcal{O}_V/f^*m(\mathbb{C}^n)$ where $m(\mathbb{C}^n)$ denotes the maximal ideal of function germs on $\mathbb{C}^n$ vanishing at 0.*

## 17  Definition of versal PDE

A *first order partial differential equation* (PDE) on $\mathbb{C}^n$ is a subvariety $V$ of the projective cotangent space $PT^*\mathbb{C}^n$ furnished with the canonical contact structure. Our subject is the local topological structure of the PDEs at the singular points of the projection of the variety to the base space $\mathbb{C}^n$. Now replace $PT^*\mathbb{C}^n$ by the 1-jet bundle $J^1(\mathbb{C}^{n-1}, \mathbb{C})$ with the contact form $\omega = dy - \sum p_i \, dx_i$, where $x, y$ are respectively the coordinates of $\mathbb{C}^{n-1}, \mathbb{C}$ and $p$ is the coordinate of $\mathbb{C}^{n-1}{}^*$, the fiber of the projection $\pi : J^1(\mathbb{C}^{n-1}, \mathbb{C}) \to \mathbb{C}^{n-1} \times \mathbb{C} = \mathbb{C}^n$.

Let $V$ be an image of an immersion $I : \mathbb{C}^n \to J^1(\mathbb{C}^{n-1}, \mathbb{C})$ such that $\pi \circ I(0) = 0 \in \mathbb{C}^{n-1} \times \mathbb{C}$. The direct image of the pull back $I^*\omega$ under the projection $\pi \circ I$ defines a multi valued 1-form (implicit differential equation) on the *configuration space* : the base space $\mathbb{C}^{n-1} \times \mathbb{C}$. A *complete integral* or *first integral* is a germ of non singular smooth function $s : \mathbb{C}^n, 0 \to \mathbb{C}, 0$ such that $ds \wedge I^*\omega$ vanishes identically on a neighbourhood of 0. Then the images $D_d = \pi \circ I(s^{-1}(d)), d \in \mathbb{C}$, constitute the integral submanifolds (possibly singular) of the equation on a neighbourhood of $0 \in \mathbb{C}^n$. We call $D_d$ a *solution* of the equation and call the family $W_I = \{D_d, d \in \mathbb{C}\}$ the *solution web* of the equation $I$.

Two webs $W_I = \{D_d\}, W_J = \{D'_d\}$ of PDEs $I, J$ are $C^\infty$ *equivalent* if there exists a germ of a contact diffeomorphism $d\psi$ of the Legendre fibration $J^1(\mathbb{C}^{n-1}, \mathbb{C}) \to \mathbb{C}^n \times \mathbb{C}$ sending the image of $I$ to that of $J$.

When $I, J$ admit first integrals $s, k$ then $I, J$ are equivalent if and only if there exist germs of diffeomorphisms $\psi : (\mathbb{C}^n, 0) \to (\mathbb{C}^n, 0)$ and $k : (\mathbb{C}, 0) \to (\mathbb{C}, 0)$ such that $\psi(D_d) = D'_{k(d)}$ for $d$ in a neighbourhood of $0 \in \mathbb{C}$. We say that $I, J$ are *strictly $C^\infty$ equivalent* if $k$ is the identity.

We say that the solution web $W_I$ of a PDE with first integral is a *non singular m-web at $q \in \mathbb{C}^n$* if the number of parameters $d$ for which $D_d$ passes through $q$ is $m$, those solutions $D_{d_i}, i = 1, ..., m$, are smooth and meet in general position at $q$ and also $D_d$ forms a non-singular foliation at $q$ as $d$ varies nearby $d_i$ for each $i = 1, ..., m$. We call the $m$ the *web number*. The web number is the topological multiplicity of $\pi \circ I$ for generic $I$. The *singular locus $Sing(W_I)$ of a web with first integral $W_I$* is the set of those $q$ where $W_I$ is not a non singular $m$-web, $m$ being the web number. Note that $Sing(W_I)$ is defined only for webs with first integrals.

An *unfolding $I'$ of dimension $r$* of a PDE $I$ is an integrable holonomic PDE on $\mathbb{C}^{n+r}$ such that

(1) The solution web $W_I$ is induced from $W_{I'}$ by the natural embedding

$\mathbb{C}^n \to \mathbb{C}^{n+r}$ "transverse" to the solutions of $I'$.

We say $I'$ is *versal* if it has the following properties.

(2) For any unfolding $I''$ on $\mathbb{C}^{n+s}$, there exists a mapping $i : \mathbb{C}^{n+s} \to \mathbb{C}^{n+r}$ such that the restriction to $\mathbb{C}^n$ is the identity and the solution web $W_{I''}$ is induced from $W_{I'}$ by $i$.

(3) Any deformation is trivial: Let $I''$ be an unfolding of $I'$. Then $I''$ is diffeomorphic to a cylinder of $I'$.

(The embedding of $\mathbb{C}^n$ is not transversal to the $W_{I'}$ in general in the ordinary sense. (3) may follow from (2).) From Condition (3) it follows that the web number of $I' \leq n+r$. In fact a mini versal unfolding (versal unfolding with minimal parameters) has web number $n + r$.

In sect. 19, 20, we proved

**Theorem 17.1** *If a holonomic PDE is smooth, generic and admits a first integral, then it admits a versal unfolding.*

The property (2) suggests that $W_{I'}$ possesses a certain universality. The classification problem of holonomic PDEs falls into the following two problems.

(A) Classification of versal PDEs $W_{I'}$;

(B) Geometry of sections of $W_{I'}$.

In the case a PDE admits a first integral, Problem A is reduced to the singularity theory of functions $s(*, q)$ on varieties $V_q$ (cf. [3,12,34,35,49,52]). Problem B is closely related to the web geometry of the solution webs, which is mentioned in the later part of this note.

## 18  From PDE with first integral to Radon transformation

Hayakawa-Ishikawa-Izumiya-Yamaguchi [37] explained a link of the first order PDEs with complete integral and the singularity theory of the so-called generating functions of Legendrian submanifolds, and classified generic PDEs with complete integrals on the plane as follows.

First recall a well known result due to Hörmander and Zakalyukin [38,68]. Let $h : \mathbb{C}^{n+k}, 0 \to \mathbb{C}^{n+1}, 0$ a germ of a smooth map of corank 1: we may assume $h(z, x) = (x, h_x(z))$, $z \in \mathbb{C}^k, x \in \mathbb{C}^n$. Assume that $(\partial h_x / \partial z_1, \dots, \partial h_x / \partial z_k) | \mathbb{C}^n \times 0$ is non singular. Then $\Sigma(h)$ is a smooth submanifold of dimension n, on which $h$ restricts to a finite-to-one and generically immersive mapping onto the discriminant set $D(h)$ assuming a certain generic condition. The Legendre submanifold associated to $h$ is the image of the map

$$(x, \ h_x(z), \ \partial h_x/\partial x_1(z), \ldots, \partial h_x/\partial x_n(z)) : \Sigma(h) \to J^1(\mathbb{C}^{n-1}, \mathbb{C}), \qquad (1)$$

which is nothing but the Nash blow up of the discriminant set $D(h)$. The family of functions $h_x$ is called the *generating function*. It is known that all germs of Legendre submanifolds of $J^1(\mathbb{C}^{n-1}, \mathbb{C})$ are obtained in this way (see e.g. [68]).

Next we recall an idea from the paper [37]. Let $I = (I_x, I_y, I_p) : \mathbb{C}^n \to J^1(\mathbb{C}^{n-1}, \mathbb{C}) = \mathbb{C}^{n-1} \times \mathbb{C} \times \mathbb{C}^{n-1*}$ be a germ of an embedding of $\mathbb{C}^n$ at 0 (not necessarily transverse to the contact elements) and assume it admits a complete integral $s$. Define the germ of Legendre embedding

$$\tilde{I} = (I_x, s, I_y, I_p, \alpha) : \mathbb{C}^n \to J^1(\mathbb{C}^n, \mathbb{C}) = \mathbb{C}^{n-1} \times \mathbb{C} \times \mathbb{C} \times \mathbb{C}^{n*}$$

with a function $\alpha$ in $\mathbb{C}^n$ such that $\tilde{I}_{\tilde{\omega}}^* = I^*\omega + \alpha \, ds = 0$, where $\omega, \tilde{\omega}$ are respectively the canonical contact forms on the jet spaces $J^1(\mathbb{C}^{n-1}, \mathbb{C}), J^1(\mathbb{C}^n, \mathbb{C})$. Then by the above construction of Legendre submanifolds, the image of $\tilde{I}$ is identified with the image of (1). We identify the divergent diagram

$$\mathbb{C} \xleftarrow{\quad s \quad} \mathbb{C}^n \xrightarrow{\quad \pi \circ I = (I_x, I_y) \quad} \mathbb{C}^{n-1} \times \mathbb{C}$$

and the restriction of the divergent diagram

$$\mathbb{C} \xleftarrow{\quad x_n \quad} \Sigma(x, h_x) \xrightarrow{\quad f = (x_1, \ldots, x_{n-1}, h_x) \quad} \mathbb{C}^{n-1} \times \mathbb{C}. \qquad (*)$$

Then the integral curve $s^{-1}(d) \subset \Sigma(x, h_x)$ is the critical point set of the restriction of $f = (x_1, \ldots, x_{n-1}, h_x) : \mathbb{C}^{n-1} \times t \times \mathbb{C}^k \to \mathbb{C}^{n-1} \times \mathbb{C}$. The solution $D_d$ is the discriminant (critical value) set of the restriction, which is the intersection of the discriminant $D(x, h_x)$ with $\mathbb{C}^{n-1} \times d \times \mathbb{C}$.

In this way we are led to the study of the divergent diagrams $(**)$ with corank $(f, x_n) = 1$,

$$\mathbb{C} \xleftarrow{\quad s = x_n \quad} \mathbb{C}^{n+k} \xrightarrow{\quad f \quad} \mathbb{C}^{n-1} \times \mathbb{C}. \qquad (**)$$

Define the *solution* $D_d$ by the discriminant set $D(f, s) \cap \mathbb{C}^n \times d$ of the restriction $f_d : s^{-1}(d) \to \mathbb{C}^n$. We denote the family of the solutions $D_d$, $d \in \mathbb{C}$ by $W_{f,s}$ and call it the *solution web* of $(f, s)$.

We proved

**Theorem 18.1** *All solution webs of holonomic PDEs with first integrals are the solution webs of divergent diagrams* $(**)$.

On the one hand the diagram $(**)$ can be regarded as a family of the restrictions of $s$ to the fibers of $f$,

$$s_q : f^{-1}(q) \longrightarrow \mathbb{C}$$

with parameter space $\mathbb{C}^n$. By definition of the solution web, $q \in D_d$ if and only if $s_q$ has the critical value $d$ for $q \in \mathbb{C}^n - D(f)$.

## 19   Thom-Mather theory

Let $\mathcal{O}(n)$ denote the local ring of the function germs on $\mathbb{C}^n$ at $0$ and $m(n)$ the maximal ideal which consists of the function germs vanishing at $0$. Similarly to the contact equivalence relation for map germs, we say divergent diagrams $(f,s) : \mathbb{C} \leftarrow \mathbb{C}^{n+k} \to \mathbb{C}^n, (g,t) : \mathbb{C} \leftarrow \mathbb{C}^{m+k} \to \mathbb{C}^m$ are *algebraically S-equivalent* if there exist a $\mathbb{C}$-algebra isomorphism $\phi^* : Q(g) \to Q(f)$ and a germ of diffeomorphism $\chi : \mathbb{C}, 0 \to \mathbb{C}, 0$ such that $\phi^*(t) = \chi \circ s$, and we say *strictly algebraically S-equivalent* if $\chi$ is the identity, where $Q(f) = \epsilon(n+k)/f * m(n)$. Roughly stating $(f,s), (g,t)$ are algebraically S-equivalent if the restrictions $s_o, t_o$ are equivalent by "coordinate change" of the source and target.

Two diagrams $(f,s), (g,t) : \mathbb{C} \leftarrow \mathbb{C}^{n+k} \to \mathbb{C}^n$ are *equivalent* if there exist germs of diffeomorphisms $\chi, \bar{\psi}, \psi$ such that the following diagram commutes

$$
\begin{array}{ccccc}
\mathbb{C}^1 & \xleftarrow{\;s\;} & \mathbb{C}^{n+k} & \xrightarrow{\;f\;} & \mathbb{C}^n \\
\chi\downarrow & & \bar{\psi}\downarrow & & \psi\downarrow \\
\mathbb{C}^1 & \xleftarrow{\;t\;} & \mathbb{C}^{n+k} & \xrightarrow[g]{} & \mathbb{C}^n.
\end{array}
$$

We denote this diagram by $(\chi, \bar{\psi}, \psi) : (f,s) \to (g,t)$ and call the triple an *equivalence*. By definition, $\psi(W_{f,s}) = W_{g,t}$ holds: $\psi(D_d) = D'_{\chi(d)}$ for $d \in \mathbb{C}$, where $D'_d$ denotes the solution of $(g,t)$. We say $(f,s), (g,t)$ are *strictly equivalent* if $\chi$ is the identity. An *unfolding of a diagram* $(f,s)$ *of dimension* $s$ is a pair of a diagram $(F,S)$ and a triple of imbeddings $i = (i_1, i_2, i_3)$ such that $i_3$ is transverse to $F = \pi \circ (F,S)$ and $(f,s)$ is given by the following commutative diagram of fiber product

$$
\begin{array}{ccccc}
\mathbb{C}^1 & \xleftarrow{\;S\;} & \mathbb{C}^{n+r+k} & \xrightarrow{\;F\;} & \mathbb{C}^{n+r} \\
\| & & i_2\uparrow & & i_3\uparrow \\
\mathbb{C}^1 & \xleftarrow[s]{} & \mathbb{C}^{n+k} & \xrightarrow[f]{} & \mathbb{C}^n.
\end{array}
$$

We denote $i : (f, s) \to (F, S)$ and call it a *morphism*.

In the manner of Thom-Mather theory we say diagrams $(f, s), (g, t)$ are (strictly) S-*equivalent* if they admit unfoldings, which are (strictly) equivalent.

Let $\theta(n)$ denote the $\epsilon(n)$-module of germs of smooth vector fields on $\mathbb{C}^n$ at 0. For a map germ $f : \mathbb{C}^{n+k}, 0 \to \mathbb{C}^n, 0$, let $\theta(f)$ denote the $\epsilon(n)$-module of germs of sections of the pull back $f^* T\mathbb{C}^n$.

We say $(f, s)$ is *(infinitesimally) strictly stable* if the morphism

$$T(f, s) : \theta(n + k) \ \oplus \ \theta(n) \to \theta(s) \oplus \theta(f)$$

defined by

$$T(f, s)(\chi, \xi) = (ts(\chi), \ tf(\chi) - \omega f(\xi))$$

is surjective, where $tf, ts$ denote the differentials of $f, s$ and $\omega f$ the pull back by $df$.

We say $f$ is *stable as a map germ* if for any deformation $g$ of $f$ a germ of $g$ at an $(z', u')$ nearby the origin is equivalent to the germ $f$. The following is classically known [30,47,48].

**Theorem 19.1** *A map germ* $f : \mathbb{C}^n, 0 \to \mathbb{C}^p, 0$ *is stable if and only if it is infinitesimally stable:* $tf - \omega f : \theta(n) \oplus \theta(p) \to \theta(f)$ *is surjective.*

The following is an easy consequence of the deformation theory of singularities of functions [53].

**Theorem 19.2** *Let* $\{(s_i, f_i), i = 1, \ldots, r\}$ *be a generator over* $\mathbb{C}$ *of the module*

$$J = \frac{\theta(s) \oplus \theta(f)}{Im \ T(f, s) + f^* m(n)(\theta(s) \oplus \theta(f))}.$$

*Define an unfolding of* $(f, s)$ $(F, S) : \mathbb{C}, 0 \leftarrow \mathbb{C}^{n+r+k}, 0 \to \mathbb{C}^{n+r}, 0$ *by*

$$F(z, u) = (f(z) + \sum_{i=1}^{r} u_i f_i, \ u), \quad S(z, u) = s(z) + \sum_{i=1}^{r} u_i s_i, \ z \in \mathbb{C}^{n+k}, \ u \in \mathbb{C}^r.$$

*Then* $(F, S)$ *is strictly stable.*

The above module $J$ is a generalization of Jacobi's ideal of functions. For example if $f$ is the projection, $J$ is the Jacobi's ideal of the restriction $s_0$ of $s$ to the fiber $(= \mathbb{C}^k)$ of $f$

$$J(s_0) = \frac{O_{\mathbb{C}^k}}{< \partial s_0 / \partial x_{n+1}, \ldots, \partial s_0 / \partial x_{n+k} >}.$$

So in general we define the *Milnor number* of $s_0$ by

$$\mu(s_0) = \dim_{\mathbb{C}} J.$$

$J$ may be regarded as the normal space of the strict equivalence class of $(f_0, s_0)$ in the space of divergent diagrams. Here $f_0 : f^{-1}(0) \to 0, s_0 : f^{-1}(0) \to \mathbb{C}$ are the restrictions of $f, s$ and $\theta(s) \oplus \theta(f)$ mod $f^* m(n)$ is the infinitesimal deformation space of $(f_0, s_0)$ and $\mathrm{Im} T(f, s)$ is the infinitesimal deformation space $(f, s)$ by coordinate change.

**Theorem 19.3** *Strictly stable diagrams possess the following properties.*

*(1) $s$ is non singular,*

*(2) $f$ is stable as a map germ,*

*(3) $(f, s) : \mathbb{C}^{n+k}, 0 \to \mathbb{C}^{n+1}, 0$ is stable as a map germ,*

*(4) the restriction $f_0 : s^{-1}(0), 0 \to \mathbb{C}^n, 0$ is stable as a map germ.*

In particular a solution web of a strictly stable divergent diagram is a one parameter family of discriminant sets of stable map germs $f_d$. The restriction $s_q$ has $m$ critical values ($m \leq n$) and the following conditions are equivalent.

*(i) The web $W_{f,s}$ is a non singular $m$-web at a $q$,*

*(ii) The first projection $\pi : D(f, s) \to \mathbb{C}^n$ is a non ramified $m$-sheet covering over $q$,*

*(iii) $s_q : f^{-1}(q) \to \mathbb{C}$ is a Morse function with $m$ distinctcritical values.*

The next theorem is fundamental in the singularity theory of composite map germs [56].

**Theorem 19.4** *Let $(f, s), (g, t) : \mathbb{C}, 0 \leftarrow \mathbb{C}^{n+k}, 0 \to \mathbb{C}^n, 0$ be strictly stable divergent diagrams. Then*

*(1) $(f, s), (g, t)$ are strictly equivalent if and only if they are strictly algebraically S-equivalent.*

*(2) $(f, s), (g, t)$ are equivalent if and only if they are algebraically S-equivalent.*

When $s_0$ admits a quasi-homogeneous ($\mathbb{C}^*$-action), conditions (1), (2) in the theorem are equivalent. Simple functions on singular varieties have been classified by Goryunov [34,39].

The theorem reduces Problem (A) to the classification of germs of functions on varieties [3,12,34,35,49,52].

The $m$-tuples of the critical values of $s_q$ define a mapping of the configuration space $\mathbb{C}^n$ onto the quotient space $\mathbb{C}^m/S(m)$ of $\mathbb{C}^m$ by the permutation group $S(m)$, $m$ being the web number. Since this map is a branched covering over the quotient space and the complement of the image of the generalized diagonal set is a $(\pi, 1)$-space, we obtain

**Theorem 19.5** *Let* $(f, s) : \mathbb{C}^{n+k}, 0 \to \mathbb{C}^{n+1}, 0 \to \mathbb{C}^n$ *be a strictly stable divergent diagram. Then the singular locus* $\mathrm{Sing}(W_{f,s})$ *of the solution web of the PDE associated to the diagram is a germ of a hypersurface. If the diagram is simple, i.e.* $s_0$ *is simple in the sense of the singularity of functions, the complement of* $\mathrm{Sing}(W_{f,s})$ *is a* $K(\pi, 1)$ *space. Here* $\pi$ *is a finite index subgroup of the braid group* $B(m)$, $m$ *being the web number, i.e. the number of the leaves passing through a generic point, and the index of* $\pi$ *is the intersection number of* $D_{d_1} \cdots D_{d_m}$ *for generic distinct* $d_1, \ldots, d_m$ *close to 0.*

Here $\mathrm{Sing}(W_{f,s})$ is the branch locus of the covering map (for the definition, see sect. 17). The above theorem suggests a relation to the ADE problem in the singularity theory.

## 20 Versality and function moduli

Let $i : (f, s) \to (F, S), (F, S) : \mathbb{C} \leftarrow \mathbb{C}^{n+\ell+k} \to \mathbb{C}^{n+\ell}$ a morphism into a strictly stable unfolding of dimension $\ell$ given in Theorem 19.2. Given a deformation $(f_v, s_v), v \in \mathbb{C}^r$ of $(f, s) = (f_o, s_o) : \mathbb{C} \leftarrow \mathbb{C}^{n+k} \to \mathbb{C}^n$, define an unfolding $(F', S')$ of $(F, S)$ with parameter $v$ by

$$F' = (F + (f_v - f, o), v), \quad S' = S + (s_v - s).$$

By Theorem 19.2, $(F', S')$ is also strictly stable and by Theorem 19.4 strictly equivalent to the trivial unfolding of $(F, S)$ of dimension $r$. The composition of the trivialization of $(F', S')$ and of the projection of the trivial family onto $(F, S)$ restricts to give a morphism $i = (i_1, i_2, i_3)$ of the subfamily $(f_v, s_v)$ into $(F, S)$. By this idea we obtain

**Theorem 20.1 (Versality theorem)** *Let* $(f_v, s_v), v \in \mathbb{C}^{r'}$ *be a smooth family of divergent diagrams and* $(F, S)$ *the versal unfolding in Theorem 19.2.*

Then $(f_v, s_v)$ is strictly equivalent to a germ of $(f'_v, s'_v)$ at an $z_v \in \mathbb{C}^{n+k}$ nearby 0, where

$$s'_v(z) = S \circ i_{2v}(z) = s(z) + \sum_{j=1}^r \alpha_{v,j}(z) \cdot s_j(z)$$

$$= s_0(z) + \sum_{j=1}^r \beta_{v,j}(f(z)) \cdot s_j(z)$$

and

$$f'_v(z) = f(z) + \sum_{j=1}^r \beta_{v,j}(f(z)) \cdot f_j(z).$$

Here the graphs of $(\alpha_{v,1}, \ldots, \alpha_{v,1}), (\beta_{v,1}, \ldots, \beta_{v,1})$ are respectively the images $i_2(\mathbb{C}^{n+k} \times v), i_3(\mathbb{C}^n \times v)$.

The second term in the theorem is called *function moduli*. It is known [30,47] that if $f$ is stable if $T(f) = tf - wf : \theta(n+k) \oplus \theta(n) \to \theta(f)$ is surjective. And then $J(f, s)$ is generated by $(s_i, 0), i = 1, \ldots, r$, where $s_i$ is the generator of $\theta(s)/ts(\ker T(f))) + f * m(n)$.

**Theorem 20.2 (Function moduli)** Let $(g, t) : \mathbb{C} \leftarrow \mathbb{C}^{m+k} \to \mathbb{C}^m$ be strictly algebraically S-equivalent to an $(f, s) : \mathbb{C} \leftarrow \mathbb{C}^{n+k} \to \mathbb{C}^n$ and assume $f, g$ are stable and $f$ is minimal: $f$ is not equivalent to a trivial unfolding of an $f'$. Then $(g, t)$ is strictly equivalent to a diagram $(F, s + s'), s' \in M$, where $F = (f, u_1, \ldots, u_{m-n})$ and $M$ is the $\epsilon(m)$-module via $F^*$ generated by $s_i, i = 1, \ldots, r$.

The second term $s' \in M$ is the function moduli.

## 21    Example (Versality and function moduli)

Consider the following (non versal) differential equation on the $x_1 y$-plane

$$y = x_1 y' + (y')^3. \tag{1}$$

This defines a nonsingular variety $V = \{y = x_1 p + p^3\} \subset J^1(\mathbb{C}, \mathbb{C})$, on which the projection $\pi$ to the base space restricts to the Whitney cusp mapping. The variety $V$ is not transverse to the contact elements at the singular locus. This equation admits the family of algebraic solutions $y = x_1 x_2 + x_2^3$, with the parameter $x_2 \in \mathbb{C}$. The variety $V$ is the image of the embedding $I(x_1, x_2) = (x_1, x_1 x_2 + x_2^3, x_2)$ and the complete integral on $V$ is given by

$s = x_2$. This admits the Legendre embedding into $J^1(\mathbb{C}^2, \mathbb{C})$ defined by $\tilde{I}(x_1, x_2) = (x_1, x_2, x_1x_2 + x_2^3, x_2, x_1 + 3x_2^2)$, which is given by the generating function $h_x(z) = z^2 + x_1x_2 + x_2^3$ in the manner of sect. 18.

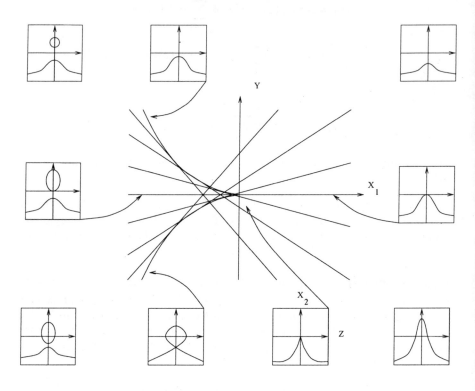

Figure 7.

Now we will construct the versal unfolding of (1). The divergent diagram

$$\mathbb{C} \xleftarrow{\ s=x_2\ } \mathbb{C}^3_{x_1 x_2 z} \xrightarrow{\ f=(x_1, h_x)\ } \mathbb{C}^2$$

is not strictly stable i.e. a deformation is not strictly equivalent to the trivial unfolding. By Theorem 19.2, this admits the stable unfolding with one parameter $u$

$$\mathbb{C} \xleftarrow{\ S=x_2\ } \mathbb{C}^4_{x_1 x_2 u z} \xrightarrow{\ F=(x_1, u, h_{x,u})\ } \mathbb{C}^3,$$

where $h_{x,u}(z) = z^2 + x_1(x_2 - u) + (x_2 - u)^3$. The singular locus $\Sigma(F, S)$ of the map $(F, S) : \mathbb{C}^4 \to \mathbb{C}^3 \times \mathbb{C}$ is the $x_1 x_2 u$-space defined by $z = 0$, on which the above divergent diagram restricts to the integral diagram

$$\mathbb{C} \xleftarrow{S=x_2} \Sigma(F,S) = \mathbb{C}^3_{x_1 x_2 u} \xrightarrow{F} \mathbb{C}^3,$$

and the restriction of $F$ is given in the coordinates $x_1, x_2, u$ by $F(x_1, x_2, u) = (x_1, u, x_1(x_2 - u) + (x_2 - u)^3)$. The level surfaces of the complete integral $s = x_2$ in $\mathbb{C}^3$ project by $F$ to the solutions in the $x_1yu$-space $\mathbb{C}^3$, which satisfy the following versal PDE

$$\left\{ \begin{array}{l} y = x_1 y_{x_1} + (y_{x_1})^3 \\ y_u = -x_1 - 3(y_{x_1})^2 \end{array} \right\}. \tag{2}$$

By definition $F^{-1}(x_1, u, y) \subset \mathbb{C}^2_{x_2 z}$ is the parallel translate of $f^{-1}(x_1, y) \subset \mathbb{C}^2$ by $(u, 0) \in \mathbb{C}^2_{x_2 z}$. Identifying $F^{-1}(x_1, u, y)$ with $f^{-1}(x_1, y)$ naturally, the restrictions $s_{x_1 y}, S_{x_1 u y}$ of the complete integral $= x_2$ satisfy $s_{x_1 y} + u = S_{x_1 u y}$ on the fiber. Let $i$ be the embedding of $x_1 y$-space into $x_1 u y$-space defined by $i(x_1, y) = (x_1, \phi(x_1, y), y)$ and $j$ the embedding of $x_1 x_2 z$-space into $x_1 x_2 u z$-space defined by $j(x_1, x_2, z) = (x_1, x_2, \phi(x_1, h_x(z)), z)$. These imbeddings define the divergent diagram $(g, t)$ by $F \circ j = i \circ g$ and $t = S \circ j$. Clearly $f = g$. By definition $S_{x_1 \phi(x_1, y) y} = s_{x_1, y} + \phi(x_1 y) : f^{-1}(x_1, y) \to \mathbb{C}$. This shows that $t = s + \phi(f)$ on $\Sigma(g, t)$. The second term $\phi(f)$ is the *function moduli*. The solutions of the equation (1) are the transverse intersections of the solutions of the equation (2) with the $x_1 y$-plane naturally embedded in the $x_1 y u$-space, and any deformation of (1) is obtained by a suitable deformation of the natural embedding.

## 22 Some normal forms of divergent diagrams

Generic PDEs with complete integrals are classified in [37,39] as follows.

**Theorem 22.1** ([37]) *The restriction of the diagrams* (*)

$$(f, s) : \mathbb{C} \leftarrow \Sigma(f, s) \to \mathbb{C}^2$$

*associated to generic differential equations in the xy-plane are equivalent to one of the following normal forms as divergent diagrams.*

$$f(z,x) = (z,x), \qquad\qquad s(z,x) = z, \qquad\qquad (0)$$
$$f(z,x) = (z^2,x), \qquad\qquad s(z,x) = z + x, \qquad\qquad (1)$$
$$f(z,x) = (z^2,x), \qquad\qquad s(z,x) = z^3 + x, \qquad\qquad (2)$$
$$f(z,x) = (z^3 + xz,x), \qquad s(z,x) = z + \phi(f), \qquad\qquad (3)$$
$$f(z,x) = (z^3 + xz,x), \qquad s(z,x) = \frac{3}{4}z^4 + \frac{1}{2}xz^2 + \phi(f), \qquad (4)$$
$$f(z,x) = (z^3 + xz^2,x), \qquad s(z,x) = z^2 + \phi(f), \qquad\qquad (5)$$

*where $\phi$ is an arbitrary smooth function defined on a neighbourhood of the origin in the xy-plane.*

The diagrams in the list correspond to the following coordinate functions on the singular curves:

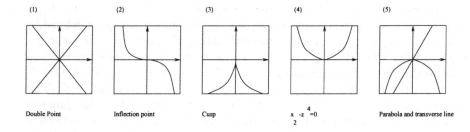

(1)     (2)     (3)     (4)     (5)

Double Point    Inflection point    Cusp    $x - \frac{z^4}{2} = 0$    Parabola and transverse line

Figure 8.

The diagrams (3), (4) and (5) correspond to non versal unfoldings of the above functions. These webs embed to versal webs $(F, S)$ on the 3-space respectively defined as follows:

$$F(z,x) = (z^3 + xz, x, y), \qquad S(z,x) = z + y, \qquad\qquad (3')$$
$$F(z,x) = (z^3 + xz, x, y), \qquad S(z,x) = \frac{3}{4}z^4 + \frac{1}{2}xz^2 + y, \qquad (4')$$
$$F(z,x) = (z^3 + xz^2, x, y), \qquad S(z,x) = z^2 + y. \qquad\qquad (5')$$

The webs in (3),(4),(5) are the transverse sections of the above versal webs by the surface $y = \phi(u, x)$ in $uxy$-space. It is important to notice that the above $(F, S)$ are the restrictions of strictly stable divergent diagrams to the singular point set $\Sigma(F, s)$, and the restrictions are not strictly stable. Some other strictly stable diagrams are seen in the papers [39,53].

In Theorem 22.1 the normal forms have at most "one" function moduli. This is explained as follows. Let $I$ be a PDE on $\mathbb{C}^n$, $I'$ its (mini) versal PDE on $\mathbb{C}^{n+s}$ with smallest $s$, and let $(f, s), (F, S)$ be their associated divergent diagrams. Define the $\mathbb{C}^+$-equivalence relation of functions germs on varieties to be generated by the strict S-equivalence and the relation $s_o \sim s_o + c, c \in \mathbb{C}$. Clearly the web structure is determined by $\mathbb{C}^+$-equivalence relation. Define the partition $S$ of the configuration space $\mathbb{C}^{n+s}$ (where the web is defined) by the $\mathbb{C}^+$-equivalence class of the singular points of the restrictions $S_{(q,u)}, (q, u) \in \mathbb{C}^{n+s}$. Assume $s_o = S_o$ is $\mathbb{C}^+$-simple: $S$ is a locally finite stratification by submanifolds, and assume that the natural embedding of $\mathbb{C}^n$ into $\mathbb{C}^{n+s}$ is transverse to $S$. The criterion in Theorem 19.2 tells us that the transversality implies that the extended family $s_{q,c} = s_q + c, (q \in \mathbb{C}^n, c \in \mathbb{C})$ is a versal family and the unfolding $(f', s'), (f' = (f, c), s' = s + c)$, of dimension 1 is strictly stable.

By the property of versal unfoldings, all deformations of $W(f, s)$ are seen on transverse sections of $W(f, s)$. Those transverse sections are presented as $c = \phi(q), q \in \mathbb{C}^n$. This argument leads us to the following theorem.

**Theorem 22.2** *Assume $(f, s)$ is $\mathbb{C}^+$-simple and generic. Then $f$ is stable and a deformation of $(f, s)$ is strictly equivalent to the normal form*

$$(f, s'), \qquad s_{q,c} = s_q + \phi(q) \quad q \in \mathbb{C}^n.$$

It would be important to estimate the codimension $c$ of the union of non $\mathbb{C}^+$-simple functions on varieties in the space of functions. For example Theorem 22.1 asserts that $c \geq 3$.

**Theorem 22.3** *For $n < c$, generic PDEs have the web number less than or equal to $n + 1$ and at most one function moduli of type $\phi(f)$.*

And also it would be interesting to classify uni-modular functions on singular varieties.

## 23  Meaning of functional moduli

The function moduli of type $\phi(f)$ has two meanings. The first one is seen by the obvious calculation

$$I_\lambda(q) = \int_{V_q} \exp \frac{i\,(s(p,q)\,+\,\phi\circ f(q))}{\lambda}\, dp$$

$$= \exp i\frac{\phi\circ f(q)}{\lambda} \times \int_{V_q} \exp \frac{i\,s(p,q)}{\lambda}\, dp.$$

The function moduli changes only the phase of the oscillatory integral. In particular the zero of the integral does not change. On the other hand, the web structure of the wave fronts $D_d$ form a $(n+1)$-web and its local structure varies topologically [55]. Such $(n+1)$-web structure may be regarded as a non-homogeneous quasi-periodic (Penrose) tiling [5]. The structure of these fine objects seems not yet well understood.

Assume that a solution web $W_I$ of a PDE $I$ on $\mathbb{C}^n$ has web number $n+1$ and has function moduli $\phi(f)$. On the complement of $\Sigma(W_I)$ the affine connection $\nabla$ as well as the web curvature form $d\theta(W_I)$ is defined, and it extends meromorphically to $\Sigma(W_I)$. Clearly the web curvature form is independent of the *right equivalence* $s \to \chi \circ s$ with a germ of diffeomorphism $\chi$ of $\mathbb{C}$. Therefore we obtain the morphism

$$K : \text{Function moduli}/ \sim^{\text{right}} \;\to\; \text{Web curvature forms.} \qquad \text{(K)}$$

## 24  Normal form of function moduli

**Theorem 24.1** *Let $(f,s)$ be the normal forms in Theorem 22.1, (3) - (5). Then it is right equivalent (respectively formally right equivalent) to the following form in case (3) (resp. in cases (4), (5)),*

*(3) $\phi$ vanishes identically on the discriminant $D(f) = \{27u^2 + 4v^3 = 0\}$,*

*(4) $\phi = v + av^2$ on the double point line $= v$-axis, $a \in \mathbb{C}$,*

*(5) $\phi = v + av^2$ on the cuspline $= v$-axis, $a \in \mathbb{C}$,*

*where $(u,v)$ are the coordinates on the target of $f$.*

The proposition is proved by normalizing by a germ of holomorphic diffeomorphism $\chi$ of $\mathbb{C}$ the dynamics on the range $\mathbb{C}$ of $s$, which sends a critical value of the function $s_q$ to the other critical value for those $q$, respectively, in the discriminant $D(f)$ and $v$-axis. The potential function $s_q$ in Case (4) for $q = (0, v)$ is of the form

$$s_q(z) = z^4 + vz^2 + \phi(v),$$

which has the critical values $\phi(v)$ and $\phi(v) + \frac{1}{4}v^2$. Assume that $\phi$ is nonsingular. Then the dynamics which sends the first to the later is conjugate to the diffeomorphism

$$v \to v + \frac{1}{4}(\phi^{-1}(v))^2 = v + v^2 - 2av^3 + \cdots.$$

It is known [46] that germs of diffeomorphisms $h$ of $\mathbb{C}, 0$ tangent to the identity are formally classified by their residues. The above dynamics has residue

$$\mathrm{res}(f) = \text{ the residue of } \frac{1}{f(v) - v} = 2a.$$

Here we introduce the (reduced) *residue*

$$\mathrm{Res}(f) = \mathrm{res}(f) + 1,$$

which may be defined by

$$f^{\circlearrowleft} = f^{(2\pi\sqrt{-1}\,\mathrm{Res}(f))},$$

in other words, writing formally as $f = \exp \chi$ with a formal vector field $\chi$ on $\mathbb{C}$,

$$\exp \circlearrowleft \chi = \exp\left(2\pi\sqrt{-1}\,\mathrm{res}(\chi)\right)\chi.$$

Here $\circlearrowleft$ stands for the "analytic" continuation of the $t$-times iteration $f^t$ of $f$, $t$ being a complex number, along a big anti clockwise cycle from 0 to a nearby point in the time space $\mathbb{C}$. The residue of our dynamics $v + \frac{1}{4}(\phi^{-1}(v))^2$ is the unique formal invariant under the right equivalence of the function moduli. It is known [26,46] that the formal conjugacy is in general divergent, to which the analysis with resurgent functions applies. Introduced in this way the residue seems an ad hoc invariant, although, the argument in sect. 10 suggests there might be an "intrinsic" definition in terms of the web curvature forms as well as the affine connections of the solutions.

The author would ask the following questions.

(C)   (i) Is $K$ injective?   (ii) Study the singularities of $K(W_I)$.

In the following sections we will see that in Case (3) the curvature forms are holomorphic in Case (4) the curvature forms have simple pole of order 2 along the cusp, and in Case (5) the curvature forms have simple pole of order 2 along $v$-axis but vanish identitically along the other irreducible components of $\Sigma(W)$. In sect. 26 we investigate some other results.

## 25    Singular 3-webs of Clairaut type: Type (3)

A *singular 3-web of Clairaut type* is a one parameter family of germs of smooth curves $D_t : (\mathbb{C}, 0) \to, (\mathbb{C}^2, p(t)), t \in \mathbb{C}$, which have contact of order 2 with the $(2,3)$-cusp $\{4x^2 + 27y^3 = 0\}$ at $p(t) = (-3c(t)^2, -2c(t)^3)$. Here $c(t)$ is a holomorphic function with $c(0) = 0, c'(0) \neq 0$ and $D_t$ is defined on a uniform neighbourhood of the origin in $\mathbb{C}$.

Now consider the Whitney mapping $f = (x, y^3 + xy)$ of $\mathbb{C}^2$. The critical point set is $\Sigma(f) = \{x = -3y^2\}$ and the discriminant set $D(f) = f\Sigma(f)$ is the $(2,3)$-cusp above presented. By the condition that $D_t$ has contact of order 2 with the cusp, $D_t$ lifts via $f$ to a smooth curve $C_t$ in the source of $f$ transverse to the critical point set $\Sigma(f)$ for small $t \neq 0$. Since the local intersection number of $C_t$ and $\Sigma(f)$ is independent of $t$, the transversality holds also at $t = 0$, and those lifts form a non singular foliation on a neighbourhood of the origin in the source of $f$. As $f$ is proper and 3-to-1, the foliation of the source project to form a 3-web on a neighourhood of the origin in the target of $f$.

**Proposition 25.1** *The family of curves $D_t$ form a 3-web structure on the complement of the cusp nearby the origin.*

The parameter $\lambda = t : \mathbb{C}^2, 0 \to \mathbb{C}, 0$ is a well defined and non singular function on the source of $f$ and the level curves form the above foliation. The *incidence surface* $I \subset \mathbb{C}^3$ is defined by

$$I = \overline{\{(u_1, u_2, u_3) \in \mathbb{C}^3 \mid D_{u_1} \cap D_{u_2} \cap D_{u_3} \neq \emptyset\}},$$

which is unique up to Cartesian product $\phi \times \phi \times \phi$ of coordinate change $\phi$ of the parameter $t$. By definition $I$ is invariant under the permutation of the coordinates of $\mathbb{C}^3$. To present $I$ in another way, identify the target space of $f$ with the quotient space of the plane $H = \{\sigma_1 = x_1 + x_2 + x_3 = 0\}$ in the $x_1 x_2 x_3$-space $\mathbb{C}^3$. The quotient map is given by $(\sigma_2, \sigma_3)$, where $\sigma_i$ denotes the symmetric polynomial of $x_1, x_2, x_3$ of degree $i$. The preimage of a point $(\sigma_2, \sigma_3)$ by $f$ is $\{(\sigma_2, x_1), (\sigma_2, x_2), (\sigma_2, x_3)\}$. Therefore the incidence set is the closure of the set of those $F = (\lambda(\sigma_2, x_1), (\lambda(\sigma_2, x_2), (\lambda(\sigma_2, x_3))$. The entries of the points $F$ are well defined on the plane $H$. By the transversality of the lift $C_t$ to the critical point set $\Sigma(f)$, the derivation $\partial\lambda/\partial y$ with respect to the second coordinate is non zero at the origin. It is easily seen that

$$\partial\lambda(\sigma_2, x_1)/\partial x_1(0, 0, 0) = \partial\lambda/\partial y(0, 0) \neq 0 .$$

Therefore the function $\lambda(\sigma_2, x_1)$ on $H$ is non singular at the origin. Since $dF$ has full rank at the origin, we obtain

**Proposition 25.2** ($^{15}$) *The incidence surface $I$ is smooth.*

On the incidence surface $I$, the coordinate functions $x_1, x_2, x_3$ induce a non singular symmetric 3-web. And the quotient of the 3-web on $I$ by the symmetry is naturally identified with the initial singular 3-web of Clairaut type. In fact the mapping

$$(u_1, u_2, u_3) : \mathbb{C}^2 \to \mathbb{C}^3$$

is a local embedding at a non singular point of the web.

By the symmetry, the incidence surface is tangent to $H$ at the origin. Let $P : I \to H$ be the linear projection such that the kernel is the strict diagonal set of $\mathbb{C}^3$. Let $P(u) = (x_1, x_2, x_3), u = (u_1, u_2, u_3) \in I$. The coordinate of $u$ is

$$(x_1 + h(\sigma_2, \sigma_3), x_2 + h(\sigma_2, \sigma_3), x_3 + h(\sigma_2, \sigma_3)),$$

where $\sigma_2, \sigma_3$ are symmetric polynomials and $\sigma_1(u) = 3h(\sigma_2, \sigma_3)$.

**Proposition 25.3** ($^{15}$) *The equivalence classes of singular 3-webs of Clairaut type correspond to the equivalence classes of the incidence surfaces under diffeomorphisms of type $\phi(u_1) \times \phi(u_2) \times \phi(u_3)$.*

*Proof.* Since the configuration space of the webs is identified with the quotient space of the incidence surfaces, an equivalence of 3-webs of Clairaut type lifts to an equivalence of incidence surfaces. On the incidence surfaces the 3-web structure is given by 3 coordinate foliations. As equivalence of incidence surfaces respects this web structure, it is of the form of Cartesian product as in the statement. On the one hand it is seen that such an equivalence of incidence surfaces induces an equivalence of singular 3-webs of Clairaut types $\square$.

**Proposition 25.4** ($^{24}$) *Incidence surfaces are equivalent by diffeomorphisms of Cartesian product to the normal forms of $I$: $I$ contains the generalized diagonal set on $H$, in other words, $h(\sigma_2, \sigma_3) = 0$ if $x_i = x_j$ for $i \neq j$*

*Proof.* The intersection of an incidence surface with the generalized diagonal set in $\mathbb{C}^3$ is a union of three curves passing though the origin, all of which are identified by symmetry. On one of those curves, two coordinates coincide, which we denote by $T$. Define a dynamics $J$ on the $T$-coordinate by $J(T) = S$ if a leaf $u_3 = $ const $T$ of the 3-web passes through a $(U, U, T)$ and also a $(S, T, T)$ on the incidence surface.

It is seen on figure 9 that $J'(0) = -2$. By Poincaré's linearization theorem there exists a germ of a diffeomorphism $\phi$ of the $T$-space at 0 which conjugates

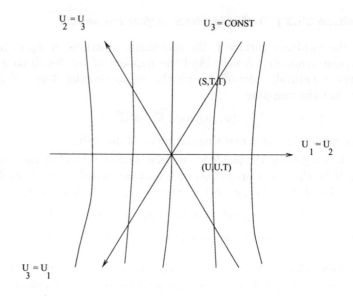

Figure 9.

$J$ and the linear map $-2T$: $-2\phi(T) = \phi \circ J(T)$. Since the dynamics $J$ is determined by the web structure, the dynamics for the image $(\phi \times \phi \times \phi)(I)$ is the linear map $-2T$. The proof is completed by the following lemma.

**Lemma 25.1** *The dynamics $J$ is linear if and only if the incidence space $I$ contains the generalized diagonal set in $H$.*

Of course, these statements can be generalized for $n + 1$-web of Clairaut type on $\mathbb{C}^n$ defined in a similar manner.

## 26 Determination by curvature form: Clairaut type (Type 3)

Let $u_i = x_i + h(\sigma_2, \sigma_3)$. The set of those $(u_1, u_2, u_3) \in \mathbb{C}^3$ with $\sigma_1 = x_1 + x_2 + x_3 = 0$ is invariant under permutation of the coordinates. The image is a germ of a hypersurface at the origin and defined by a symmetric function $f(\Sigma_1, \Sigma_2, \Sigma_3)$, where $\Sigma_i$ is a symmetric polynomial of $u_1, u_2, u_3$ of degree $i$. By the symmetry, the quotient of the surface is smooth and transverse to the $X$-axis in the quotient space of $u_1 u_2 u_3$-space with coordinates $X = \Sigma_1, Y = \Sigma_2, Z = \Sigma_3$, and may be presented by $X - H(Y, Z) = 0$, and then

$f = \Sigma_1 - H(\Sigma_2, \Sigma_3)$. Both the coordinates $(x_1, x_2, x_3)$ and $(u_1, u_2, u_3)$ define the same generalized diagonal set in $x_1 x_2 x_3$-space.

**Proposition 26.1** *There is a one-to-one correspondence of the k-jets of $h(\sigma_2, \sigma_3)$ with respect to $x_1, x_2, x_3$ and the k-jets of $H(\Sigma_2, \Sigma_3)$ with respect to $u_1, u_2, u_3$.*

*Proof.* Clearly the $k$-jet of the image of the plane $\{\sigma_1 = 0\}$ under $(u_1, u_2, u_3)$ is determined by that of $h(\sigma_2, \sigma_3)$. On the one hand, the image is presented by the web function $\Sigma_1 - H(\Sigma_2, \Sigma_3) = 0$. Therefore the $k$-jet of the image is determined by that of $H$ in the coordinates $(u_1, u_2, u_3)$ $\square$.

By the argument in sect. 6, the web curvature form is

$$d\theta = \frac{1}{2} \sum_{i \neq j} \frac{\partial^2}{\partial u_i \partial u_j} \log \frac{f_i}{f_j} du_i \wedge du_j = \frac{1}{2} \sum_{i \neq j} \frac{f_{iij}}{f_i^2} - \frac{f_{ijj}}{f_j^2} + f_{ij} \left( \frac{f_{ii}}{f_i^2} - \frac{f_{jj}}{f_j^2} \right) du_i \wedge du_j$$

Write the curvature form as

$$d\theta = k \, dx_1 \wedge dx_2 == k \, dx_2 \wedge dx_3 = k \, dx_3 \wedge dx_1$$

then $f$ is skew-invariant under $S(3)$-action and written as

$$k = -(x_1 - x_2)(x_2 - x_3)(x_3 - x_1) \, k'(\sigma_2, \sigma_3).$$

Since

$$-(x_1 - x_2)(x_2 - x_3)(x_3 - x_1) dx_i \wedge dx_{i+1} = d\sigma_2 \wedge d\sigma_3$$

the curvature form induces the following web curvature form on the quotient space

$$k'(\sigma_2, \sigma_3) \, d\sigma_2 \wedge d\sigma_3.$$

We prove

**Theorem 26.1** *Let $H = (u_1 - u_2)^2 (u_2 - u_3)^2 (u_3 - u_1)^2 H'(\Sigma_2, \Sigma_3)$. Then the formal operator from $H'$ to $k'$ is bijective.*

*Proof.* Denote

$$H = \sum_{d=0}^{\infty} \sum_{2\alpha + 3\beta = d} a_{\alpha\beta} \, \Sigma_2^{\alpha} \Sigma_3^{\beta}$$

$$f' = \sum_{d=0}^{\infty} \sum_{2\alpha' + 3\beta' = d} b_{\alpha'\beta'} \, \Sigma_2^{\alpha'} \Sigma_3^{\beta'}$$

By the definitions of $\Omega, f', b_{\alpha'\beta'}$ is a polynomial of those $a_{\alpha\beta}$ with

$$2\alpha + 3\beta \leq 2\alpha' + 3\beta' + 6.$$

Let $P_{\alpha\beta}$ denote the sum of those monomials $b_{\alpha'\beta'}\sigma_2^{\alpha'}\sigma_3^{\beta'}$ of degree $2\alpha + 3\beta - 6$ with $2\alpha' + 3\beta' = 2\alpha + 3\beta - 6$ and $a_{\alpha\beta}$ in their coefficients $b_{\alpha'\beta'}$. ($P_{\alpha\beta}$ is the lowest order contribution of $a_{\alpha\beta}$ to the web curvature form.) Clearly those terms appear in the Taylor expansion of

$$\frac{1}{2}\sum_{i\neq j}\frac{f_{iij}}{f_i^2} - \frac{f_{ijj}}{f_j^2}du_i \wedge du_j.$$

Since $f_i(0) = 1$ and $du_i = dx_i + dh(\sigma_2, \sigma_3)$ for $i = 1, 2, 3$, the coefficients of $P_{\alpha\beta}$ appear in

$$\sum_{i<j}\{(a_{\alpha\beta}\Sigma_2^\alpha\Sigma_3^\beta)_{iij} - (a_{\alpha\beta}\Sigma_2^\alpha\Sigma_3^\beta)_{ijj}\}du_i \wedge du_j.$$

Since $du_i \wedge du_j = dx_i \wedge dx_j + (dx_j - dx_i) \wedge dh(\sigma_2, \sigma_3)$ and $dx_i \wedge dx_j$ is independent of $i < j$, we see by direct calculation the lowest order contribution of $a_{\alpha,\beta}$ in the above term is

$$(\frac{\partial^3}{\partial X^3} + 2\frac{\partial^2}{\partial Y^2} + \sigma_2\frac{\partial^3}{\partial X\partial Y^2} + \sigma_3\frac{\partial^3}{\partial Y^3})a_{\alpha\beta}\sigma_2^\alpha\sigma_3^\beta$$
$$\times (x_1 - x_2)(x_2 - x_3)(x_3 - x_1) \, dx_1 \wedge dx_2$$

where $X, Y$ stand for $\sigma_2, \sigma_3$. Therefore

$$P_{\alpha\beta} = -a_{\alpha\beta}(h_{\alpha\beta}\sigma_2^{\alpha-3}\sigma_3^\beta + k_{\alpha\beta}\sigma_2^\alpha\sigma_3^{\beta-2})$$

and

$$h_{\alpha\beta} = \alpha(\alpha - 1)(\alpha - 2)$$
$$k_{\alpha\beta} = \beta(\beta - 1)(\alpha + \beta).$$

Now replace $H$ with

$$(u_1 - u_2)^2(u_2 - u_3)^2(u_3 - u_1)^2 H' = (4\Sigma_2^3 + 27\Sigma_3^2)H',$$

which vanishes identically on the generalized diagonal set $\Delta$. Let $Q_{\alpha\beta}$ denote the sum of the lowest order terms of the curvature form on the quotient space $\Omega$ of degree $d = 2\alpha + 3\beta$ with $a_{\alpha\beta}$ in the coefficients. Then

$$Q_{\alpha\beta} = -4a_{\alpha\beta}(h_{\alpha+3,\beta}\sigma_2^\alpha\sigma_3^\beta + k_{\alpha+3\beta}\sigma_2^{\alpha+3}\sigma_3^{\beta-2})$$
$$- 27a_{\alpha\beta}(h_{\alpha\beta+2}\sigma_2^{\alpha-3}\sigma_3^{\beta+2} + k_{\alpha\beta+2}\sigma_2^\alpha\sigma_3^\beta) \, .$$

Let $H_d$ denote the (weighted) homogeneous polynomial part of $H'$ of degree $d$

$$H_d = \sum_{2\alpha+3\beta=d} a_{\alpha\beta}\Sigma_2^\alpha\Sigma_3^\beta$$

and let $K_d$ denote the (weighted) homogeneous polynomial part of the curvature $k'$ of degree $d$

$$K_d = \sum_{2\alpha+3\beta=d} P_{\alpha\beta} \quad (= \sum_{2\alpha+3\beta=d} c_{\alpha\beta}\sigma_2^\alpha\sigma_3^\beta).$$

Then the linear operator which sends $H_d$ to $K_d$ is given by the matrix

$$\begin{pmatrix} k_{\alpha_1,\beta_1+2}\, h_{\alpha_1+3,\beta_1} & 0 & \cdots 0 \\ 0 & k_{\alpha_2,\beta_2+2}\, h_{\alpha_2+3,\beta_2} & \cdots 0 \\ & & \\ 0 & 0\cdots & k_{\alpha_\ell,\beta_\ell+2}\, h_{\alpha_\ell+3,\beta_\ell} \end{pmatrix} \begin{pmatrix} 27 & 0 & \cdots & 0 \\ 4 & 27 & \cdots & 0 \\ 0 & 4 & \cdots & 0 \\ \cdots & \cdots & & \\ 0 & \cdots & 4 & 27 \\ 0 & \cdots & 0 & 4 \end{pmatrix}$$

in the basis $\sigma_2^{\alpha_i}\sigma_3^{\beta_i}$ and $\Sigma_2^{\alpha_i}\Sigma_3^{\beta_i}$ for $i = 1,\dots,\ell$ respectively in the source and the target, where $(\alpha_i,\beta_i)$ is the pair of non negative integers such that $2\alpha_i + 3\beta_i = d$ and $\alpha_i$ is increasing of $i$. By a theorem of Sylvester, it is seen that this matrix is of full rank. This completes the proof of the proposition$\Box$.

We may restate the result as follows.

**Theorem 26.2** *The following spaces are formally equivalent.*

*Moduli space of singular 3-webs of Clairaut type*

$$\simeq \frac{germs\ of\ holomorphic\ 2\text{-}forms\ on\ \mathbb{C}^2}{\mathbb{C}^*\text{-}action\ with\ weight\ (2,3)preserving\ the\ cusp}$$

$\simeq$ *germ of holomorphic 2-forms $\Omega$ on $\mathbb{C}^2$ such that*

$$\Omega = \sum_{d=d_0}^\infty \sum_{3i+2j=d} a_{ij}x^iy^j\ dx \wedge dy, \quad \sum_{3i+2j=d_0} a_{ij} = 1.$$

## 27 Remark on the formal equivalence of moduli and curvature forms

In order to show that the above formal equivalences give actually a bijection of those spaces, it suffices to see the convergence of the formal power series of the web equation for a given a holomorphic curvature 2-form. Here it seems that the problem of small denominator does not arise. However in this note we show the following partial results.

**Proposition 27.1** *Let* $W = (F_1, F_2, F_3), W' = (F_1', F_2', F_3')$ *be 3-webs of normal form in Sect. 7, i.e.* $F_2 = F_2' : dx$ *and* $F_3 = F_3' : dy$. *Assume that their curvature forms coincide with each other. Then* $W = W'$.

*Proof.* Let $f, g$ be defining functions of $F_1, F_1'$. Then the curvature forms of $W, W'$ are

$$d\theta(W) = \frac{\partial^2}{\partial x \partial y} \log \frac{f_x}{f_y} \ dx \wedge dy$$

$$d\theta(W') = \frac{\partial^2}{\partial x \partial y} \log \frac{g_x}{g_y} \ dx \wedge dy.$$

By the assumption,

$$\frac{f_x}{f_y} = \frac{g_x}{g_y} \ A(x)B(y)$$

holds with meromorphic functions $A, B$. Now let $C : \mathbb{C}^2, (0,0) \to \mathbb{C}^2, (0,0)$ be a germ of a diffeomorphism such that

$$g(C(s,t)) = t, \quad C(-t,t) = (0,t), \quad C(0,t) = (t/2, t/2), \quad C(t,t) = (t,0),$$

for all small $s, t$, and consider the function $f(C(s,t)) - t$ of $s$ parameterized by $t$.

We may assume that $f(C(s,0))$ is not identically 0, in other words, the transverse leaves $f = 0, g = 0$ do not coincide. By the normalization conditions in sect. 7,

$$f(C(s,t)) = 0 \quad \text{for} \quad s = -t, 0, t.$$

Assume $s = -t, 0, t$ have multiplicities $d_1, d_2, d_3$ respectively. Then the family $f(C(s,t) - t$ is a deformation of 0 with multiplicity $d \geq d_1 + d_2 + d_3$. Such a family has (possibly multiple) $d - 1$ critical points nearby $s = 0$ for small $t$. Those points $s = -t, 0, t$ have multiplicities $d_1 - 1, d_2 - 1, d_3 - 1$. Therefore there exist more than 2 critical points different from $-t, 0, t$. Let $D$ denote the set of those points where the level curves of $f, g$ have contact of order $\geq 2$. It is seen that the curve coincides with the set of $C(s,t)$, such that $s$ is a critical function of $f(C(s,t)) - t$. Therefore the curve $D$ splits into a union of the $y$-axis with multiplicity $d_1 - 1$, the diagonal set with multiplicity $d_2 - 1$, the $x$-axis with multiplicity $d_3 - 1$ and a curve $D'$ with multiplicity 2 at the origin.

On $D'$, $A(x)B(y) = 1$ holds. It is seen that $D'$ is singular at the origin, and then it follows that $A(x), B(y)$ are constant. Therefore

$$\frac{f_x}{f_y} = \frac{g_x}{g_y} A(x)B(y) = \frac{g_x}{g_y}$$

holds identically on a neighbourhood of the origin $\square$.

The statement remains valid also for the real $C^3$-case.

## 28 Realization of curvature forms by singular web structure

The following proposition tells that all holomorphic 2-forms are realized as web curvature forms of singular 3-webs of Clairaut type in sect. 25.

**Proposition 28.1** *Given a germ of a holomorphic 2-form $\Omega$, there exits a singular 3-web of Clairaut type with the 2-form $\Omega$ as its web curvature form.*

*Proof.* Let $\tilde{\Omega}$ be the lift of $\Omega$ to the plane $H = \sigma_1 = 0$ in $xyz$-space. By Poincaré's lemma there exists a $\theta'$ such that $d\theta' = \tilde{\Omega}$. Clearly $\tilde{\Omega}$ is invariant under the permutation of the coordinates $x, y, z$ ($x = x_1, y = x_2, z = x_3$). Define $\tilde{\theta}$ by the average under the permutation group $S(3)$

$$\tilde{\theta} = \frac{1}{6} \sum_{\sigma \in S(3)} \sigma^* \theta'.$$

Then $\tilde{\theta}$ is also $S(3)$-invariant and $d\tilde{\theta} = \tilde{\Omega}$. Let $f$ a function such that $\tilde{\theta} - f(dx + dy)$ is closed, and let $d\log g = \theta - f(dx + dy)$. Then $g$ is a non zero and holomorphic on a neighbourhood of the origin. Define

$$\omega_1' = g\,(dx + dy)$$

and

$$\omega_1'' = \frac{1}{2}(\omega_1' + T^*\omega_1')$$

where $T$ is the involution transposing the coordinates $x, y$. Then

$$d\,\omega_1' = \tilde{\theta} \wedge \omega_1', \quad \text{and} \quad d\,\omega_1'' = \tilde{\theta} \wedge \omega_1''.$$

Let $R$ be the generator of the index 2 subgroup, which sends $(x, y, z)$ to $(z, x, y)$. Define

$$\omega_2'' = R^*\omega_1''$$

and

$$\omega_3'' = -\omega_1'' - \omega_2''$$

and define

$$\tilde{\omega}_i = \frac{1}{3}(\omega_i'' + R^* \omega_{i+1}'' + (R^*)^2 \omega_{i+2}'')$$

where the index $i$ is an integer mod 3. Then $\tilde{\omega}_i$ is non singular, invariant under the transposition of the coordinates $x_i, x_{i+1}$ and

$$(R^j)^* \tilde{\omega}_i = \tilde{\omega}_{i+j} \tag{1}$$

for all $i, j$, where $x_1, x_2, x_3$ stand for $x, y, z$ respectively. By construction

$$\tilde{\omega}_1 + \tilde{\omega}_2 + \tilde{\omega}_3 = 0$$

and

$$d\tilde{\omega}_i = \tilde{\theta} \wedge \tilde{\omega}_i$$

holds for $i = 1, 2, 3$. By (1), $\tilde{\omega}_i, i = 1, 2, 3$ induces an equal 3-web structure on the quotient space of Clairaut type. Clearly the 3-web of $\tilde{\omega}_1, \tilde{\omega}_2, \tilde{\omega}_3$ has the curvature form $d\tilde{\theta} = \tilde{\Omega}$ and therefore the quotient has the prescribed curvature form $\Omega$. $\qquad\square$

The 3-web of Clairaut type constructed as above is not necessarily of the normal form in Theorem 24.1 and Proposition 25.4. However starting with a suitable foliation instead of $dx + dy$, we may realize the curvature form by Clairaut type of the normal form.

Finally the author would ask the following question. Assume two holonomic PDEs $V, V' \subset J^1(\mathbb{C}^{n-1}, \mathbb{C})$ with the same singular locus in $\mathbb{C}^{n-1} \times \mathbb{C}$ have the same connection $\nabla$ as a multi valued affine connection on $\mathbb{C}^{n-1} \times \mathbb{C}$.

(D) What can one state about $V$ and $V'$? Or in the same vein, which part of the singularities of the connection $\nabla$ remains invariant under variation of function moduli in Theorem 20.1, 20.2 ?

## References

1. M. A. AKIVIS & A. M. SHELEKHOV, *Geometry and algebra of multi-codimensional three-webs*, Kluwer Academic Publ., 1992.
2. V. I. ARNOLD, *Critical points of smooth functions*, Proc. ICM (Vancouver) **1** (1974), 19–41.
3. V. I. ARNOLD, *Wave front evolution and equivariant Morse lemma*, Comunication Pure Appl. Math. **29** (1976), 557–582.

4. V. I. ARNOLD, *Contact geometry and wave propagation*, L'Enseignement Math. Supplement, 1989.

5. V. I. ARNOLD, *Huygens and Barrow, Newton and Hooke*, Birkhäuser, 1990.

6. V. I. ARNOLD, A. N. VARCHENKO, A.B. GIVENTAL & A.G. HOV-ANSKY, *Singularities of functions, wave fronts, caustics and multidimensional integrals*, Math. Phys. Rev. **4** (1984), 1–92.

7. V. I. ARNOL'D, S. M. GUSEIN-ZADE & V. N. VARCHENKO, *Singularities of differentiable mappings*, Monographs in Math. 1, Birkhauser, 1985.

8. I. N. BERNSTEIN, *Modules over a ring of differential operators. An investigation of the fundamental solutions of equations with constant coefficients*, Funct. Anal. Appl. **5** (1971), no. 2, 1–16.

9. W. BLASCHKE, *Einführung in die Geometrie der Waben*, Birkhäuser, Basel-Stuttgart, 1955.

10. W. BLASCHKE & G. BOL, *Geometrie der Gewebe*, Grundlehren der Mathematischen Wissenschaften, Springer-Verlag, 1938.

11. J. B. BRUCE, *A note on first order differential equations of degree greater than one and wave front evolutions*, Bull. London Math. Soc. **16** (1984), 139–144.

12. J. BRUCE & M. ROBERT, *Critical points of functions on analytic varieties*, Topology 27 (1988), 57–90.

13. R. L. BRYANT, *Notes on the paper "Abel's Theorem and Webs" by S.S. Chern and P.A. Griffiths*, non published.

14. R. L. BRYANT, S. S. CHERN, R. B. GARDNER, H. L. GOLDSCHMIDT & P. A. GRIFFITHS, *Exterior Differential Systems*, Mathematical Science Research Institute Publications, vol. 18, Springer-Verlag, 1991.

15. M.J.D. CARNEIRO, *Singularities of envelopes of families of subvarieties in* $\mathbb{R}^n$, Ann. Sci. Ecole Norm. Sup. **16** (1983), no. 2, 173–192.

16. D. CERVEAU, *Equations différentielles algébriques: remarques et problèmes*, J. Fac. Sci. Univ. Tokyo, Sect. IA Math. **36** (1989), no. 3, 665–680.

17. D. CERVEAU, *Théorème de type Fuchs pour les tissus feuilletés*, Complex analytic methods in dynamical systems (Rio de Janeiro, 1992), Astérisque, no. 222, 1994.

18. S. S. CHERN, *Abzählungen für Gewebe*, Abh. Hamburg **11** (1936), 163–170.

19. S. S. CHERN, *Web geometry*, Bull A.M.S. **6** (1982), 1–8.

20. J. DAMON, *The unfolding and determinacy theorems for subgroups of* $\mathcal{A}, \mathcal{K}$, Memoirs A.M.S. 50, 1984.

21. L. DARA, *Singularités génériques des équations différentielles*, Bol. Soc. Brasil Mat. (1975), 95–127

22. G. DARBOUX, *Théorie des surfaces*, t. 1 (1914), 151–161.

23. J. P. DUFOUR, *Triplets de fonctions et stabilité des enveloppes*, C. R. Acad. Sci. Paris **10** (1977), 509–512.

24. J. P. DUFOUR, *Modules pour les familles de courbes planes*, Ann. Inst. Fourier **39** (1989), no. 1, 225–238.

25. A. DU PLESSIS, *Genericity and smooth finite determinacy*, Proc. AMS Symp. Pure Math. 40, 1 (1983), 295–312.

26. J. ECALLE, *Théorie itérative. Introduction à la théorie des invariants holomorphes*, J. Math. Pures Appl. **54** (1975), 183–258.

27. M. V. FEDORYUK, *Asymptotic analysis*, Springer, 1991.

28. I. M. GELFAND & I. ZAKHAREVICH, *Webs, Veronese curves and bihamiltonian systems*, J. Funct. Anal. **99** (1991), 150–178.

29. E. GHYS, *Flots transversalement affines et tissus feuilletés*, Mem. Soc. Math. France **46** (1991), 123–150.

30. C. G. GIBSON ET AL. *Topological stability of smooth mappings*, Lecture Notes in Math. 552, Springer-Verlag, 1976.

31. V. V. GOLDBERG *Theory of multicodimensional $(n + 1)$-webs*, Kluwer Academic Publ., Mathematics and Its Applications, vol. 44, 1988.

32. V. V. GOLDBERG, *Maks Aizikovich Akivis*, Tver University, Russia, (1993), 4–8.

33. V. V. GOLDBERG, *Gerrit Bol (1906-1989) and his Contribution to web geometry*, in Webs and Quasigroups, Tver University, Russia, 1994.

34. V. V. GORYUNOV, *Geometry of bifurcation diagram of simple projections onto the line*, Funct. Anal. Prilozhen. **14** (1981), no. 2, 1–8.

35. V. V. GORYUNOV, *Projection and vector fields, tangent to the discriminant of a complete intersection*, Funct. Anal. Prilozhen. **22** (1988), no. 2, 26–37.

36. F. GUILLEMIN & S. STERNBERG, *Geometric Asymptotics*, Mathematical surveys, vol. 14, AMS, 1977.

37. A. HAYAKAWA, G. ISHIKAWA, S. IZUMIYA & K. YAMAGUCHI, *Classification of integral diagrams of Whitney type and first order differential equations*, Internat. J. Math. **5** (1994), 447–489.

38. L. HÖRMANDER, *Fourier Integral Operators.I*, Acta Mathematica **127** (1971), 79–183.

39. S. IZUMIYA, *Completely integrable holonomic systems of first-order differential equations*, Proc. Roy. Soc. Edinburgh Sect. A **125** (1995), no. 3, 567–586.

40. J. L. JOLY, G. MÉTIVIER & J. RAUCH, *Resonant one dimensional nonlinear geometric optics*, J. Funct. Anal. **114** (1993), 106–231.

41. J. L. JOLY, G. MÉTIVIER & J. RAUCH, *Trilinear compensated compactness and nonlinear geometric optics*, Ann. Math. **142** (1995), 121–169.

42. M. KOSSOWSKI, *First order partial differential equations with singular solution*, Indiana Univ. Math. Jour. **35** (1986), 209–222.

43. Y. KUROKAWA, *On function moduli for first order ordinary differential equations*, C. R. Acad. Sci. Paris (1993).

44. S. LIE, *Bestimmung aller Flächen, die in mehrfacher Weise durch Translationsbewegung einer Kurve erzeugt werden*, Ges. Abhandlungen. Bd. **1**, 450-467; Arch. für Math. Bd. **7** Heft 2 (1882), 155–176.

45. S. LIE, *Das Abelsche Theorem und die Translationsmannigfaltigkeiten*, Leipziger Berichte (1897), 181–248; Ges. Abhandlungen. Bd.II, Teil II, paper XIV, 580-639.

46. B. MALGRANGE, *Travaux d'Ecalle et de Martinet-Ramis sur les systèmes dynamiques*, Séminaire Bourbaki, exposé 582, 1981.

47. J. N. MATHER, *Stability of $C^\infty$-mappings: I-VI*, Ann. Math.**87** (1968), 89–104; **89** (1969), 254–291; Publ. IHES **35** (1969), 127–156, **37** (1970), 223–248; Adv. Math.**4** (1970), 301–335; Lecture Notes in Math.**192** (1971), 207–253.

48. J. N. MATHER, *Stratifications and mappings*, in Salvador Symposium on Dynamical systems (ed. M. Peixoto), Academic Press, New York (1973), 195–232.

49. S. MATSUOKA, *An algebraic criterion for right-left equivalence of holomorphic functions on analytic varieties*, Bull. London Math. Soc. **21** (1989), 164–170.

50. K. MAYRHOFER, *Kurven systeme auf Flächen*, Math. Z. **29** (1928), 728–752.

51. K. MAYRHOFER, *Uber sechsen systeme*, Abh. Math. Sem. Univ. Hamburg **7** (1929), 1–10.

52. D. MOND & J. MONTALDI, *Deformations of maps on complete intersections, Damon's $\mathcal{K}_v$-equivalence and bifurcations*, Warwick University Preprints 21, 1991.

53. I. NAKAI, *Versal unfolding of PDE with complete integrals and web structure of solutions*, J. Diff Equations **118** (1995), no. 2, 253–292.

54. I. NAKAI, *$C^\infty$-stability and the $I_0$-equivalence of diagrams of smooth mappings*, preprint, Liverpool University, 1986.

55. I. NAKAI, *Topology of complex webs of codimension one and geometry of projective space curves*, Topology **26** (1987), 147–171.

56. I. NAKAI, *Topological stability theorem for composite mappings*, Ann. Inst. Fourier **39** (1989), no. 2, 459–500.

57. I. NAKAI, *Nice dimensions for the $I_0$ equivalence relation of diagrams of map germs*, Pacific J. Math. **147** (1991), no. 2, 325–353.

58. I. NAKAI, *Superintegrable foliations and web structure*, in Geometry and Analysis in Dynamical systems (ed. H. Ito), Advanced Series in Dynamical Systems 14, World Scientific, (1994), 126–139.

59. I. NAKAI, *Curvature of curvilinear 4-webs and pencils of one forms: variation on a theorem of Poincaré, Mayrhofer and Reidemeister*, Coment. Math. Helv. **73** (1998), no. 2, 177–205.

60. T. PEARCEY, *The structure of an electromagnetic field in the neighbourhood of a cusp of a caustic*, 1945.

61. H. POINCARÉ, *Sur les surfaces de translation et les fonctions abéliennes*, Bul. Soc. Math. France **29** (1901), 61–86; in Oeuvres, t. VI, 13-37.

62. F. PHAM, *Résurgence d'un thème de Huygens-Fresnel*, IHES Publ. Math. **68** (1988), 77–90.

63. K. REIDEMEISTER, *Gewebe und Gruppen*, Math. Z. **29** (1928), 427–435.

64. H. SATO & A. YOSHIKAWA, *Third order ordinary differential equations and Legendre connections*, J. Math. Soc. Japan 50 (1998), no. 4, 993-1013.

65. S. TABACHINIKOV, *Geometry of Lagrangian and Legendrian 2-web*, Differential Geom. Appl. **3** (1993), no. 3, 265–284.

66. R. THOM, *Sur les équations différentielles multiformes et leurs intégrales singulières*, Bol. Soc. Brasil Mat. **3** (1972), no. 1, 1–11.

67. A. M. VINOGRADOV, *Many-valued solutions and a classification principle for differential equations* Soviet Math. Dokl. **14** (1972), 661–665.

68. V. M. ZAKALYUKIN, *Lagrangian and Legendrian singularities*, Funct. Anal. Appl. **10** (1976), 37–45.

# VERONESE WEBS AND TRANSVERSALLY VERONESE FOLIATIONS

MARIE-HÉLÈNE RIGAL

*Ecole Normale Supérieure, 46, Allée d'Italie, 69364 LYON Cedex 7,*
*Marie-Helene.RIGAL@umpa.ens-lyon.fr*

## 1 Introduction

Three foliations of dimension 1 determine a web on a surface $S$ if they are transverse to each other at each point. These structures are rigid and are characterized by local invariants [2].

Webs can also be considered as structures transverse to some foliations. Let us consider the following case :

**Definition 1** [5]. A foliated web on a 3-manifold is a flow $\mathcal{F}$ which is the intersection of three codimension one foliations $\mathcal{F}^0, \mathcal{F}^1, \mathcal{F}^\infty$ that are transverse to each other.

*Examples :*

1.1 Let $M$ be a locally trivial fibration by circles over the torus $\mathbb{T}^2$ endowed with a web. The foliation $\mathcal{F}$ whose leaves are the fibers of the fibration is a foliated web.

1.2 Let $A$ be a hyperbolic isomorphism in $SL_2(\mathbb{Z})$ (Tr $A > 2$). It has two eigenvalues. When considered as an isomorphism of the torus $\mathbb{T}^2$ it is then associated with two linear flows on $\mathbb{T}^2$. Let $\mathbb{T}^3_A$ be the suspension of $A$ over the circle $S^1$. This manifold is a bundle whose fibre is a torus and whose basis is the circle. The proper directions of $A$ then define two foliated webs on $\mathbb{T}^3_A$.

1.3 Let the manifold $M^3_\lambda$ be the quotient of $\mathbb{R}^3 \setminus \{0\}$ by the group generated by the homothety of positive ratio $\lambda$, $\lambda \neq 1$ : $M^3_\lambda$ is diffeomorphic to the Hopf manifold $S^1 \times S^2$. The foliation of $\mathbb{R}^3 \setminus \{0\}$ by vertical lines projects onto a foliated web on $M^3_\lambda$.

The transverse rigidity of foliated webs implies that the examples above are (up to conjugation) the only possible cases of foliated webs on closed connected 3-manifolds [5].

In fact foliated webs appear naturally in different situations [5,4]. They namely coincide with bihamiltonian structures in dimension 3 and the preceding examples 1.1 and 1.2 are precisely the bihamiltonian systems on closed 3-manifolds.

More generally, the study of bihamiltonian systems in odd dimension $2n + 1$ leads to the study of foliations of codimension $n + 1$ whose transverse structure is the following generalization of the usual notion of a web in dimension 2 :

**Definition 2.** *A* conformally parallelizable (**CP**) *structure on an n-dimensional manifold M corresponds to $n + 1$ foliations $\mathcal{F}^0, \cdots, \mathcal{F}^n$ of dimension 1 such that the associated directions are transverse to each other and span the entire tangent space $T_x M$ at every point x.*

The aim of this paper is to present some elementary facts concerning these structures and their rigidity and to describe the foliations that have a transverse CP structure. The transversally CP flows will then be classified and a precise description of bihamiltonian systems in odd dimension will be given.

All the manifolds and foliations will be supposed to be smooth and orientable.

## 2 Conformally parallelizable structures - Examples

Let $\mathcal{F}^0, \cdots, \mathcal{F}^n$ be a CP structure on the $n$-dimensional manifold $M$. Definition 2 is equivalent to say that every point $x$ has a neighborhood on which $n$ vector fields $(v_0, \cdots, v_{n-1})$ satisfying the following property are defined :

$$v_i(x) \text{ is tangent to } \mathcal{F}^i \text{ and } v_0 + \cdots + v_{n-1} \text{ is tangent to } \mathcal{F}^n. \qquad (*)$$

Actually, $(v_0, \cdots, v_n)$ is a projective frame of $T_x M$. Consequently, the notion of direction has an intrinsic and global meaning on $M$ : $D = (a_0, \cdots, a_n)$ is the direction associated to the 1-form $\omega^D = a_0 \omega^0 + \cdots + a_n \omega^n$. The local system characterized by $(*)$ and defined up to multiplication by a positive function will be called *adapted frame* for the CP structure. The dual basis of $(v_0, \cdots, v_{n-1})$ will be noted $(\omega^0, \cdots, \omega^{n-1})$.

*Examples* :

2.1 The webs on surfaces are the CP structures in dimension 2.

2.2 Let $(x_1, \cdots, x_n)$ be the usual coordinates of $\mathbb{R}^n$. The canonical CP structure of $\mathbb{R}^n$ is given by $(dx_1, \cdots, dx_n)$

**Definition 3.** *A CP structure is* flat *if it is locally the same as the canonical CP structure on* $\mathbb{R}^n$.

Since the canonical CP structure of $\mathbb{R}^n$ is invariant by translations, the torus $\mathbb{T}^n$ can also be endowed with a canonical flat CP structure.

2.3 Let the manifold $M_\lambda^n$ be the quotient of $\mathbb{R}^n \backslash \{0\}$ by the group generated by the homothety of positive ratio $\lambda$, $\lambda \neq 1$ : $M_\lambda^n$ is diffeomorphic to the Hopf manifold $S^1 \times S^{n-1}$. The CP structure of $\mathbb{R}^n \backslash \{0\}$ is given by $(\frac{1}{r} dx_1, \cdots, \frac{1}{r} dx_n)$, where $r$ is the module of the vector $(x_1, \cdots, x_n)$. It projects onto a CP structure on $M_\lambda^n$. This is nothing but an affine structure on $S^1 \times S^{n-1}$ whose holonomy is formed by homotheties. With the preceding example on the torus $\mathbb{T}^n$, they are the only possible cases of CP flat structures on an $n$-dimensional closed manifold [3].

2.4 *Veronese webs.*

**Definition 4** [4]. *A* Veronese web *on an n-dimensional manifold is a CP structure such that each local adapted frame* $(\omega^0, \cdots, \omega^{n-1})$ *satisfies :*

$$\forall t \in \mathbb{R} \cup \{\infty\}, \qquad \omega_t = \omega^0 + t\omega^1 + \cdots + t^{n-1}\omega^{n-1} \quad \text{is integrable.}$$

The integrability condition $\omega_t \wedge d\omega_t = 0$ means that for every $t$, $\omega_t$ is associated with a codimension 1 foliation $\mathcal{F}_t$.

2.4.1 All the flat CP structures are Veronese webs.

2.4.2 *The canonical Veronese web on* $SL_2(\mathbb{R})$. The Lie group $SL_2(\mathbb{R})$ is the group of $2 \times 2$ matrices of determinant 1. Its Lie algebra $sl_2$ is identified with the $2 \times 2$ matrices of trace 0. The Killing form on $sl_2$ is given by the determinant. It is a quadratic form of signature $(1, 2)$. Let $\mathcal{C}$ be its light cone (the set of isotropic vectors) and $(\omega^0, \omega^1, \omega^2)$ be the left invariant 1-forms on $SL_2(\mathbb{R})$ associated to the dual basis of the following basis of $sl_2$:

$$\sigma_0 = \frac{1}{2} \begin{pmatrix} 0 & 1 \\ 0 & 0 \end{pmatrix}, \quad \sigma_1 = \frac{1}{2} \begin{pmatrix} 1 & 0 \\ 0 & -1 \end{pmatrix}, \quad \sigma_2 = \frac{1}{2} \begin{pmatrix} 0 & 0 \\ -1 & 0 \end{pmatrix}$$

One easily checks that $(\omega^0, \omega^1, \omega^2)$ is a Veronese web on $SL_2(\mathbb{R})$. The planes tangent to the light cone $\mathcal{C}$ in $sl_2$ are exactly the kernels of the $\omega_t$ at the identity. This also induces a canonical Veronese web on the universal covering $\widetilde{SL_2(\mathbb{R})}$ of $SL_2(\mathbb{R})$.

# 3 Rigidity of CP structures

The word rigid structure will signify that the group of diffeomorphisms which preserve a CP structure is a Lie group of finite dimension. The language and usual tools of $G$-structures will be used to prove this result.

## 3.1 Webs in dimension 2

Let us recall that a web is locally determined by three 1-forms $\omega^0, \omega^1, \omega^\infty$ such that $\omega^1 = \omega^0 + \omega^\infty$. Of course these forms are integrable and a 1-form $\gamma$ is completely determined by the following equations :

$$d\omega^0 = \gamma \wedge \omega^0 \qquad d\omega^1 = \gamma \wedge \omega^1 \qquad d\omega^\infty = \gamma \wedge \omega^\infty$$

This is the connection associated to the web. The curvature $d\gamma$ of $\gamma$ is an invariant of the web which is flat if and only if $d\gamma$ is equal to zero.

## 3.2 CP structures in higher dimensions

Let us recall that the frame bundle of an $n$-dimensional manifold $M$ is a principal bundle $B \xrightarrow{\pi} M$ with structural group $GL_n(\mathbb{R})$. A point $z_x$ of $E$ over $x$ is a frame of $T_x M$, i.e a linear isomorphism $z_x : \mathbb{R}^n \to T_x M$.

From the definition of a CP structure in dimension $n$ it is easy to see that the set of adapted frames forms a principal subbundle $E$ of $B$. Its structure group is isomorphic to $(\mathbb{R}_+^*, \times)$ since it is made of the positive homotheties of $\mathbb{R}^n$. Let $z_x$ be a point of $E$. The vertical space $V_{z_x}$ in $z_x$ is the subspace of $T_{z_x} E$ which is tangent to the action of $\mathbb{R}_+^*$. A (linear) connection on $E$ is nothing but a differentiable distribution of hyperplanes of TE (the *horizontal* distribution) supplementary to $V$ and invariant under the action of $\mathbb{R}_+^*$. Equivalently, a connection on $E$ is a one form $\omega$ which satisfies the following condition :

$$s'^* \omega = s^* \omega + \frac{df}{f} \qquad f \in C^\infty(M)$$

where $s' = fs$ and $s$ are sections of $\pi : E \to M$.

The curvature of $\omega$ is the two form $d\omega$.

The torsion of $\omega$ is a two form $T$ with values in $\mathbb{R}^n$. It is defined by

$$\forall i = 0, \cdots, n-1, \qquad T^i = d\omega^i + s^* \omega \wedge \omega^i$$

where $s = (\omega^0, \cdots, \omega^{n-1})$ is a local section of $E$.

**Definition 5.** *The canonical connection $\omega$ associated to the CP structure is the unique one form such that for every $i$ (mod $n$)*

$$T^i \wedge \omega^0 \wedge \cdots \wedge \hat{\omega}^i \wedge \hat{\omega}^{i+1} \wedge \cdots \wedge \omega^n = 0.$$

It is easy to check that the one form $\omega$ defined above is indeed a connection.

**Proposition 1.** *A CP structure is flat if and only if its canonical connection $\omega$ is torsionless. In this case the CP structure corresponds to an affine structure on the manifold and its holonomy is made of homotheties.*

*Proof.* The assertion will be proved in dimension 3. The generalization to higher dimensions is straightforward.

When the CP structure is flat it is clear that the canonical connection also is flat (it has neither torsion nor curvature).

On the other hand let us write in a local section $s = (\omega^0, \omega^1, \omega^2)$ of $E$ that $\omega$ is torsionless $j = 0, 1, 2$ :

$$d\omega^j + s^*\omega \wedge \omega^j = 0$$

Thus $s^*\omega$ is closed which means that it is locally exact. It is possible to suppose that $s^*\omega = df$ if the open subset on which $s$ is defined is small enough. The forms $e^{-f}\omega^j$ are closed. Consequently, the CP structure is flat.

Finally, as the connection is flat it is associated to an affine structure. $\square$

**Proposition 2.** *If the torsion of $\omega$ is not identically zero, a canonical conformal class of riemannian metrics is defined on an open set of $M$.*

*Proof.* It will be proved in dimension 3. The generalization in every dimension is easy.

Let us note that the torsion of the canonical connection is equal to zero on a neighborhood of a point $x$ in $M$ if and only if for every direction $D = (a_0, a_1, a_2)$ and every local adapted frame $(\omega^0, \omega^1, \omega^2)$ the forms $\omega^D = a_0\omega^0 + a_1\omega^1 + a_2\omega^2$ are integrable.

Let $x$ be a point of $M$ in which the torsion does not vanish. Let us fix a direction $D$ such that whatever the local adapted frame $\omega^D \wedge d\omega^D$ does not vanish around $x$. The set of points $y$ such that $\omega^D_y \wedge d\omega^D_y$ is non zero is open. Let $\Omega^D$ be the connected component containing $x$. Around every point $y$ in $\Omega^D$ and for every local frame, it is possible to write $\omega^D \wedge d\omega^D = \varphi^D \cdot \omega^0 \wedge \omega^1 \wedge \omega^2$. Then $g^D = (\varphi^D)^2 \left( (\omega^0)^2 + (\omega^1)^2 + (\omega^2)^2 \right)$ is a well defined metric. Its conformal class does not depend on the direction $D$ and is defined on $\Omega = \bigcup_D \Omega^D$. □

*Remark.* The choice of a representative in the conformal class of Riemannian metrics allows us to define a canonical parallelism over $\Omega$ by choosing at each point the unique adapted frame whose first vector is unitary for this metric.

The notion of a geodesic passing through a point $x$ of $M$ with speed $v_x$ can be defined in the same way as in riemannian geometry. Let us recall this construction and some properties in the case of a CP structure.

If $v_x$ is a tangent vector in $T_x M$ it determines a direction $D_{v_x}$ on $M$. Let $z_x$ be a point of $E$ such that $\pi(z_x) = x$. There is a unique horizontal vector $X^D$ defined on $E$ by the relations

$$\pi_*(X^D_{z_x}) = v_x \qquad \pi_*(X^D_{z_y}) = z_y \circ z_x^{-1}(v_x)$$

Clearly $\pi_*(X^D_{z_y})$ is tangent to the direction $D_{v_x}$. The geodesic passing through $x$ with speed $v_x$ is the projection of the orbit of $X^D$ passing through $z_x$. Thus a geodesic has a constant direction. It is worth noting that $X^D$ may be non complete since $E$ is not compact. Consequently, the canonical connection $\omega$ has no reason for being geodesically complete. Let $\varphi_t^{v_x}$ be the local one parameter group of $X^D$. The exponential $exp_x$ at $x$ is then the map from a neighborhood of $0$ in $T_x M$ to a neighborhood of $x$ in $M$ which is defined by

$$exp_x(v) = \pi(\varphi_1^{v_x}(x))$$

It is a diffeomorphism if restricted to a small enough neighborhood of $0$. The study of CP structure can thus be locally linearized. As a consequence of the existence of the exponential map, an automorphism of a CP structure is entirely determined by its value at a point and the value of its differential at this point.

Besides, the Lie algebra of the vector fields which preserve the CP structure is of finite dimension at most $n+1$. The group $Aut(M)$ of diffeomorphisms which preserve a CP structure is then a Lie group of finite dimension at most equal to $n + 1$ [8]. The CP structure is *transitive* if $Aut(M)$ acts transitively on $M$.

**Proposition 3.** *In dimension 3 the only non affine and transitive Veronese web on a connected and simply connected manifold is the canonical Veronese web on* $\widetilde{SL_2(\mathbb{R})}$.

Thus the 3-dimensional manifolds endowed with a transitive and non affine Veronese web are the locally homogeneous spaces modelled on $SL_2(\mathbb{R})$.

*Proof.* As the Veronese web is transitive its torsion is either identically zero or everywhere non zero. The proposition corresponds to the second case. Let us fix a parallelism of $M$ by assuming the normalization condition

$$\omega^1 \wedge d\omega^1 = 4\omega^0 \wedge \omega^1 \wedge \omega^2$$

It is straightforward to write the conditions $\omega_t \wedge d\omega_t = 0$ for every $t$ and to check that the form $\gamma = (\omega^0, \omega^1, \omega^2)$ takes its values in $sl_2$ and satisfies the integrability condition $d\gamma + \frac{1}{2}[\gamma, \gamma] = 0$. A theorem of Darboux leads to the result. □

## 4 Transversally conformally parallelizable (TCP) foliations

In this section the CP structures will be considered as transverse structures to some foliations. Actually, the preceding sections can be considered as the particular cases of TCP foliations by points.

### 4.1 Definitions and examples of TCP foliations

A foliation $\mathcal{F}$ of codimension $p$ on an $n$ dimension manifold $M$ corresponds to a family $(U_i, \pi_i)_{i \in I}$ such that :

- $(U_i)_{i \in I}$ is a covering of $M$ by connected and simply connected open subsets of $M$.

- $\forall i \in I, \pi_i : U_i \to \mathbb{R}^p$ is a submersion. The open set $\pi_i(U_i)$ is called a local basis for $\mathcal{F}$.

- if $U_i \cap U_j \neq \emptyset$, $\varphi_{ij} = \pi_i \circ \pi_j^{-1}$ is a diffeomorphism between $\pi_j(U_i \cap U_j)$ and $\pi_i(U_i \cap U_j)$

The leaves of $\mathcal{F}$ are the embedded connected submanifolds whose restrictions to every $U_i$ coincide with the fibers of $\pi_i$.

In the same way as an $n$-dimensional manifold consists in gluing open subsets diffeomorphic to $\mathbb{R}^n$, an $n$-dimensional manifold with a foliation $\mathcal{F}$ of codimension $p$ consists in gluing trivial fibrations diffeomorphic to $\mathbb{R}^{n-p} \times \mathbb{R}^n$.

*Examples of foliated manifolds $(M, \mathcal{F})$* :

4.1 $M$ is a locally trivial fibration, $\mathcal{F}$ is the foliation whose leaves are the fibers of the fibration. These foliations are said to be *simple*.

4.2 The one dimensional foliation by parallel lines on $\mathbb{R}^n \backslash \{0\}$ induces a one dimensional foliation on $M_\lambda^n$.

Let us identify a local basis with a local transversal to $\mathcal{F}$. Local diffeomorphisms between local transversals are obtained by making these transversals slide along pathes included in the leaves. The germs of the diffeomorphisms built in such a way depend only on the homotopy class (in the leaves) of the path. The pseudo-group of holonomy $\mathcal{P}$ of $\mathcal{F}$ is formed by all these diffeomorphisms. The (group of) holonomy of a leaf $F$ is then the group of germs of diffeomorphisms of $\mathcal{P}$ which fix a point $x$ chosen on $F$. The holonomy can also be considered as a group homomorphism between the fundamental group of the leaf and the group of diffeomorphisms of $\mathbb{R}^p$ [6].

*Examples :*

- A simple foliation has a trivial holonomy.

- In the case of the flow on $M_\lambda^n$, the two non compact leaves have a non trivial holonomy which is the group generated by the homothety of ratio $\lambda$. The other leaves are non compact and simply connected. Thus their holonomy is trivial.

*4.2 Transverse geometry*

The foliation $\mathcal{F}$ has a transverse structure when the transition maps $\varphi_{ij}$ preserve a geometric structure on $\mathbb{R}^p$.

*Examples :*

- If all the $\varphi_{ij}$ are affine transformations, $\mathcal{F}$ is said to be transversally affine. In this case the holonomy coincides with the linear holonomy. For example, the flow on $M_\lambda^n$ is a transversally affine flow whose holonomy is formed by homotheties. Another example can be given with flows of codimension 2 whose pseudo-group of holonomy is made of affine maps $f : \mathbb{C} \simeq \mathbb{R}^2 \to \mathbb{C}$ such that $f(z) = az + b$, $a$ and $b$ in $\mathbb{C}$. Such flows are called transversally affine complex [5].

- If the $\varphi_{ij}$ are isometries which preserve a riemannian metric on $\mathbb{R}^p$, $\mathcal{F}$ is said to be a riemannian foliation.

- If the $\varphi_{ij}$ preserve a parallelism on $\mathbb{R}^p$, $\mathcal{F}$ is said to be transversally parallelizable. A linear flow on a torus is an example of such a foliation. The global geometry of transversally parallelizable flows on closed manifolds is well known [11] : all the leaves are diffeomorphic, their closures are submanifolds and are the fibers of a locally trivial fibration. When $\mathcal{F}$ is a flow, they are tori on which $\mathcal{F}$ is conjugated to a linear flow.

**Definition.** *A TCP foliation is a foliation with a CP transverse structure.*

*Examples :*

4.3 The transversally parallelizable foliations are TCP foliations. On the other hand let us remark that a TCP foliation is transversally parallelizable if and only if it is riemannian : the transverse parallelism is then chosen as being at each point the only adapted transverse frame with unitary first vector.

4.4 A locally trivial fibration over a manifold endowed with a CP structure. For example, a fibration by circles above $M_\lambda^n$ or $\mathbb{T}^n$ with their canonical CP structures are TCP foliations.

4.5 The flow on $M_\lambda^n$ is a TCP foliation.

Of course, a TCP foliation whose transverse CP structure is flat is a transversally affine foliation whose holonomy is made of homotheties.

**Definition 7.** *A foliated form with respect to $\mathcal{F}$ is a differential form whose restriction to every $U_i$ is identified with the pull-back by $\pi_i$ of a form on the local basis $\mathbb{R}^p$.*

*A vector field is foliated if it is projectable by $\pi_i$ on each local basis.*

*Let $x$ be a point of $M$. A transverse frame at $x$ is a frame of $T_xM/T_x\mathcal{F}$. The set of these frames is the bundle of transverse frames.*

In the case of TCP foliations the transverse frame bundle can be reduced to the $\mathbb{R}_+^*$-subbundle $E_T$ of transverse and adapted frames. The foliation $\mathcal{F}$ lifts to a foliation $\mathcal{F}_T$ of same dimension whose leaves are coverings of leaves of $\mathcal{F}$ [11]. Exactly in the same way as for CP structures, a canonical connection $\omega_T$ can be defined on $E_T$. By construction of $\omega_T$, $\mathcal{F}_T$ is horizontal (i.e $\omega_{T|_{\mathcal{F}_T}} = 0$) and $\omega_T$ is foliated for $\mathcal{F}_T$. This connection has a torsion and the propositions 1 and 2 are still valid : if the torsion is everywhere zero $\mathcal{F}$ is transversally affine, if it does not vanish at a point, $\mathcal{F}$ is transversally parallelizable on an open subset of $M$ saturated by the leaves of $\mathcal{F}$. The existence of $\omega_T$ implies the

**Proposition 4.** *The holonomy of a TCP foliation is determined by its linear holonomy which is formed by homotheties.*

*Proof.* By definition, the holonomy is made of diffeomorphisms which preserve the transverse CP structure and fix a point. Thus they are determined by their differential (the linear holonomy) which is an homothety. $\qquad \square$

*In the sequel, the manifolds endowed with TCP foliations will be supposed to be connected and closed.*

**Theorem 1** [13]. *Every TCP foliation $\mathcal{F}$ whose transverse CP structure is not flat is transversally parallelizable for an adapted transverse parallelism.*

*Sketch of the proof of theorem 1.*

As the canonical connection $\omega_T$ is invariant under the action of $\mathbb{R}^*_+$ and is foliated for $\mathcal{F}_T$, the subset where its torsion does not vanish is an open saturated subset of $E_T$ which projects onto an open subset $\Omega$ of $E$ saturated by $\mathcal{F}$. Proposition 2 and the example 5.3 imply that $\mathcal{F}$ is transversally parallelizable on $\Omega$. Two steps are then necessarily

1. Firstly one proves that $\Omega$ also contains the closures of the leaves of $\mathcal{F}_{|\Omega}$. Thus they are submanifolds of $M$ [15].

2. Secondly one proves that the boundary $\partial \Omega$ is empty. This last step uses the existence of a connection, of geodesics which have a constant direction and of the exponential map. The main difficulty is that the canonical connection is not complete.

Actually, the theorem which is proved is that every TCP foliation which is transversally parallelizable (for a parallelism adapted to its transverse CP structure) on an open subset of $M$ saturated by the leaves of $\mathcal{F}$ and their closures is in fact transversally parallelizable on $M$.

**Corollary** (*global stability of $\mathcal{F}$*). *If a TCP foliation has a compact leaf without holonomy all the leaves are compact and without holonomy.*

In this case, $\mathcal{F}$ is a simple foliation.

*Proof.* The open subset $\Omega$ of $M$ made of compact leaves without holonomy is not empty. Actually $\mathcal{F}_{|\Omega}$ is a simple foliation $\pi : \Omega \to W$ [6]. The transverse CP structure of $\mathcal{F}$ projects onto a CP structure on $W$. An adapted transverse parallelism to $\mathcal{F}$ is then determined by the choice of an arbitrary metric on $W$. $\qquad \square$

# 5 A first application : classification of TCP flows

**Proposition 5.** *Examples 4.3, 4.4, 4.5 are the only possible cases of TCP flows (up to conjugation).*

*Proof.* Let $\mathcal{F}$ denote a TCP flow on an $n$-dimensional manifold $M$. The preceding theorem implies that it is transversally parallelizable or transversally affine. The last situation will be studied here. One has to prove that examples 4.3, 4.4, 4.5 are the only possible cases of homotheties-translation flows. The ideas are due to E. Ghys for transversally affine complex flows of codimension 2 [5]. They can be extended in higher dimensions in our case. This generalization will be sketched in this section.

Let us suppose that $\mathcal{F}$ is not transversally parallelizable. According to the corollary, if there is one compact leaf without holonomy, $M$ is a circle fibration over an $(n-1)$-dimensional manifold endowed with a flat CP structure by projection of the transverse structure of $\mathcal{F}$. This corresponds to example 4.4.

One can suppose that the compact leaves (if any) have a non trivial holonomy, thus a contracting holonomy in this case. Compact leaves are then isolated and in finite number since $M$ is compact. Let us suppose the existence of $p$ compact leaves $F_1, \cdots, F_p$. They are contained in $p$ disjoint compact neighborhoods $V_1, \cdots, V_p$ that are diffeomorphic to $S^1 \times D^{n-1}$, in which $\mathcal{F}$ is conjugated to the suspension of the closed ball $D^{n-1}$ by an homothety of positive ratio $\lambda_i$ and is transverse to the boundary $\partial V_i \simeq S^1 \times S^{n-2}$. In particular the transverse flat CP structure of $\mathcal{F}$ induces an affine structure on these boundaries which allows to identify $\partial V_i$ with $M_{\lambda_i}$. Let us note $N = M \backslash (\bigcup\limits_{i=1}^{p} \text{Int } V_i)$. The foliation $\mathcal{F}_{|N}$ restricted to $N$ is transverse to $\partial N$ and is still a flat TCP foliation. The leaves of $\mathcal{F}_{|N}$ meet at most two components of $\partial N$. If a leaf of $\mathcal{F}_{|N}$ meets one component $\partial V_i$ of $\partial N$ it is proper since it meets it in a unique point.

Let us suppose that $\mathcal{F}_{|N}$ has a non compact leaf $F$. It is also possible to suppose that it is not proper and that its $\omega$-limit set $\Lambda$ is contained in the interior Int $N$ of $N$. Indeed if $F$ is a non compact leaf whose $\omega$-limit set meets $\partial N$ one sees that $F$ also meets $\partial N$. Therefore $F$ is a proper leaf. As it is relatively compact its closure contains a minimal set. As $N$ does not contain leaves diffeomorphic to circles, this minimal set is either an exceptional leaf or a compact leaf that meets two different components of $\partial N$. In the last case $F$ necessarily coincides with this compact leaf which is impossible. In the first case the exceptional leaf is contained in Int $N$. So is its $\omega$-limit set. It is then possible to construct a measure $\mu$ which is transverse to the flow, ergodic and whose support is contained in $\Lambda$. This measure has no atom. One key point in [5] is that the elements of the pseudo-group of holonomy $\mathcal{P}$ are diffeomorphisms that send balls of $\mathbb{R}^2$ on balls of $\mathbb{R}^2$. Of course this property is satisfied by homotheties in every dimension. A straightforward generalization of [5] allows us to prove that under this hypothesis, $\mathcal{P}$ satisfies a property of equicontinuity which implies that $\mathcal{F}_{|_N}$ actually is transversally parallelizable. Consequently, all the leaves of $\mathcal{F}_{|_N}$ are diffeomorphic. They are all proper since it is the case for the ones which meet $\partial N$. Which contradicts the hypothesis. Thus the leaves of $\mathcal{F}_{|_N}$ are all compact and $(N, \mathcal{F})$ is diffeomorphic to the manifold $S^{n-1} \times [0,1]$ endowed with the simple vertical foliation by intervals $[0,1]$. This corresponds to example 4.5.

If $\mathcal{F}$ has no compact leaf, then the arguments concerning the pseudo-group $\mathcal{P}$ still work and $\mathcal{F}$ is transversally parallelizable. $\qquad\square$

## 6 A second application : bihamiltonian systems in odd dimension

### 6.1 Bihamiltonian systems

*The case of even dimensions.*

Among dynamical systems, hamiltonian systems correspond to the natural setting for studying the dynamics of physical and mechanical systems.

A hamiltonian system is a symplectic manifold $(M, \omega)$ with a function $H$ (corresponding to the energy of the system). Let us recall that a symplectic form $\omega$ on a manifold $M$ is a closed 2-form which is non degenerate, i.e at each point $x$, $\omega_x : T_x^*M \to T_xM$ is a linear isomorphism. In particular the dimension of $M$ is necessary even. The dual tensor of $\omega$ is nothing but a Poisson tensor of maximal rank on $M$. A typical example of a symplectic manifold is given by the cotangent space $T^*M$ of a manifold $M$. The canonical symplectic form on $T^*M$ is the exterior derivative of the Liouville form $\theta$ defined by $\theta_{\gamma_x}(v_\gamma) = \gamma_x(\pi_* \gamma)$ where $\pi$ is the projection of $T^*M$ onto $M$, $\gamma_x$ lies in $T_x^*M$ and $v_\gamma$ in $T_\gamma(T^*M)$.

The problem is to determine the trajectories of the hamiltonian gradient $X_H$ which is the vector field defined by the equation $i_{X_H}\omega = dH$. When $n$ functions $f_1, \cdots, f_n$ which are constant along the trajectories of $X_H$ (such functions are first integral of the system) can be found so as to have independent differentials on an open dense subset and to be in involution ($[X_{f_i}, X_{f_j}] = 0$), it is possible to solve explicitly this equation (at least on an open dense subset of $M^{2n}$). The system is then said to be *completely inte-*

grable [9],[?]. In this case the geometry is well known : according to the Arnold Liouville theorem, this open subset is a bundle whose fibers are lagrangian tori of dimension $n$. The gradient $X_H$ is tangent to them and its restriction to each fiber is conjugated to a linear flow.

However a lot of interesting hamiltonian systems are not integrable. For example let $S$ be a surface imbedded in $\mathbb{R}^3$ and $g$ be the riemannian metric induced on $S$ by the euclidean structure of $\mathbb{R}^3$. The geodesic flow on the cotangent space $T^*S$ of $S$ is the hamiltonian flow $X_g$ on $T^*M$. It describes the free movement of a solid on $S$. When the curvature of $(S,g)$ is negative, the system is not integrable [7]. In fact, it is usually not easy to determine if a hamiltonian system is integrable or not. Bihamiltonian systems are then interesting as they give many examples of completely integrable systems [10].

*The case of odd dimensions.*

The quotient along the trajectories of a vector field which preserves the symplectic form (or the restriction to a transversal of this vector field if its orbits are not closed) leads to the study of analogous systems in odd dimension in the general context of Poisson structures [17].

Let us recall that a *Poisson structure* on a manifold $M$ is a skew symmetric $\mathbb{R}$-bilinear map $\{,\} : C^\infty(M) \times C^\infty(M) \to C^\infty(M)$ such that :

- For all $f$ the map $X_f : g \mapsto \{f,g\}$ is a derivation which is identified with a vector field.

- $X_{\{f,g\}} = [X_f, X_g]$.

The *Poisson tensor* $\Pi : T^*M \to TM$ is defined by $\Pi(df) = X_f$. It is skew symmetric. Its rank is even and when it is maximal in each point $\Pi$ is said to be of *maximal rank*. In this case, if $M$ is an even dimensional manifold, $\Pi$ is the dual tensor of a symplectic form. If the dimension of $M$ is odd, $Im\Pi$ is an integrable distribution in the tangent bundle $TM$. It is then associated to a codimension 1 foliation $\mathcal{S}$. By construction, the leaves $S$ of $\mathcal{S}$ are symplectic manifolds for the restriction $\Pi_{|S}$ of $\Pi$.

It is worth noting that the sum of two Poisson tensors generally is not a Poisson tensor.

**Definition 8** [4]. *A smooth non constant function $H$ and two Poisson tensors $\Pi^0$, $\Pi^\infty$ on a manifold $M$ of dimension $2n+1$ form a* bihamiltonian system in odd dimension $(M^{2n+1}, \Pi^0, \Pi^\infty, H)$ *if they satisfy :*

- *Every linear combination $\Pi^t = \Pi^0 - t\Pi^\infty$ is a Poisson tensor.*

- *The hamiltonian gradient $X_H^0$ of $H$ with respect to $\Pi^0$ is non zero and preserves $\Pi^\infty$.*

Let $\Pi^0$ and $\Pi^\infty$ be two Poisson tensors satisfying the properties of the above definition. For every $x$, their restrictions to the tangent space $T_x M$ are two skew symmetric bilinear forms of maximal rank. It is the same for every linear combination $\Pi_x^t = \Pi_x^0 - t\Pi_x^\infty$. It is then an exercice of linear algebra to prove the existence of a basis $(e_0^x, \cdots, e_n^x, f_1^x, \cdots, f_n^x)$ of $T_x M$ such that [4]

$$\Pi_x^0 = \sum_{i=1}^n e_i^x \wedge f_i^x, \qquad \Pi_x^0 = \sum_{i=1}^n e_{i-1}^x \wedge f_i^x.$$

The kernel of $\Pi_x^t$ is the linear form $e_x^0 + te_x^1 + \cdots + t^n e_x^n$ of $T_x^* M$ where $(e_x^0, \cdots, e_x^n, f_x^1, \cdots, f_x^n)$ is the dual basis of $(e_0^x, \cdots, e_n^x, f_1, \cdots, f_n^x)$. In particular, $(e_x^0, \cdots, e_x^n)$ is unique up to homothety.

This is still locally true on $M$. In other words, every point has a neighborhood on which it is possible to define $n+1$ 1-forms $(\omega^0, \cdots, \omega^n)$ such that for all $t$

$$\omega_t = \omega^0 + t\omega^1 + \cdots + t^n \omega^n$$

is integrable and is associated to the symplectic foliation $\mathcal{S}^t$. Consequently the intersection $\mathcal{A}$ of all the $\mathcal{S}^t$ is nothing but the intersection of $n+1$ different $\mathcal{S}^t$. Thus it is a foliation of codimension $n+1$. It will be called the *axis* of the bihamiltonian system. By construction the one forms $(\omega^0, \omega^1, \cdots, \omega^n)$ are foliated with respect to $\mathcal{A}$. They are canonically defined up to multiplication by a function. This means that the transverse structure of $\mathcal{A}$ is a particular case of CP structure. Precisely it is a transverse Veronese web.

## 6.2 Global geometry in odd dimension

**Theorem 2** [14]. *The axis $\mathcal{A}$ of a bihamiltonian system $(M^{2n+1}, \Pi^0, \Pi^\infty, H)$ on a closed manifold of dimension $2n+1$ is a transversally parallelizable foliation.*

In particular, when all the leaves of $\mathcal{A}$ are compact, the situation is analogous in odd dimension to the complete integrability in even dimension.

*Proof.* Firstly it will be proved that $H$ is constant along the leaves of $\mathcal{A}$. It is equivalent to prove the

**Lemma 1.** *The hamiltonian gradient $X_H^0$ is tangent to $\mathcal{A}$.*

*Proof.* The hamiltonian gradient $X_H^0$ is non identically zero. As it preserves $\Pi^t$ for every $t$, it preserves every symplectic foliation $\mathcal{S}^t$. Thus it preserves their intersection $\mathcal{A}$. If one writes that $\mathcal{L}_{X_H^0}\omega_t = \lambda(t)\omega_t$ one sees that $\lambda$ does not depend on $t$. Consequently $X_H^0$ also preserves the transverse Veronese web structure, thus its CP transverse structure.

By construction $X_H^0$ is tangent to $\mathcal{S}^0$. Let $z$ be one of its critical points (which exist since $M$ is compact). Let $\mathcal{T}_z$ be a local and connected transversal to $\mathcal{A}$ containing $z$. The transverse CP structure of $\mathcal{A}$ projects on $\mathcal{T}_z$ to a CP structure. Let $\overline{X_H^0}$ be the projection of $X_H^0$ on $\mathcal{T}_z$. It preserves the CP structure of $\mathcal{T}_z$ and fixes $z$. Its local one parameter group is then formed by some diffeomorphisms whose differential in $z$ are homotheties. Thus $\overline{X_H^0}$ is radial. It is also tangent to the direction determined by $\mathcal{S}^0$, it vanishes on a neighborhood of $z$. The lemma follows since $M$ is connected. $\qquad\square$

**Lemma 2.** *There exists an open subset saturated by the leaves and the closures of the leaves of $\mathcal{A}$ on which $\mathcal{A}$ is transversally parallelizable for a parallelism adapted to its CP transverse structure.*

*Proof.* Let $\Omega$ be the open set made of the regular points of the function $H$. It is non empty and saturated by the leaves of $\mathcal{A}$ since $H$ is basic for $\mathcal{A}$. If $x$ is a point of $\Omega$ let $(v^0, \cdots, v^n)$ be the unique transverse adapted frame such that

$$|dH|^2 = v^0(H)^2 + \cdots + v^n(H)^2$$

is equal to 1. This is a transverse and adapted parallelism for $\mathcal{A}$ on $\Omega$. By choosing a regular value $c$ of $H$ one sees that $\Omega$ also contains the closures of the leaves which are included in the compact subset $H^{-1}(c)$. Thus it contains the closures of all the leaves of $\mathcal{A}$ that are already in $\Omega$. $\qquad\square$

Theorem 2 is then a consequence of theorem 1. $\qquad\square$

From this theorem it is not difficult to check that the two first examples of foliated webs can be realized as bihamiltonian systems in dimension 3. Bihamiltonian systems in dimension 5 can also be described precisely [14]. However this is not as straightforward as in dimension 3.

## 7 Open problems

Concerning the classification of TCP flows, it could be interesting to study the flows whose pseudo-group of holonomy is made of affine maps which are similitudes or even conformal maps. The classification has been made in codimension 2 [5]. However there only are partial results in higher codimensions [1,12].

Concerning bihamiltonian systems, the regularity condition (all the $\Pi^t$ are of maximal rank) seems to be very strong.

Let us consider the following example on the Lie-Poisson group $SL_2(\mathbb{R})$ : $\mathcal{H}$ is the Lie subalgebra of $sl_2$ made of the upper triangular matrices whose trace equals zero. Let us fix a basis of $sl_2$ such that the two first vectors lie in $\mathcal{H}$. Let $\omega$ be an area form on $\mathcal{H}$. It induces a left invariant 2-form $\overline{\omega}$ and a right invariant 2-form $\tilde{\omega}$ on $SL_2(\mathbb{R})$. The dual tensors $\overline{\Pi}$ and $\tilde{\Pi}$ are Poisson tensors of maximal rank and are compatible : $\Pi^t = \overline{\Pi} - t\tilde{\Pi}$ is a Poisson tensor. However $\Pi^1$ vanishes at the identity since $\overline{\Pi}$ and $\tilde{\Pi}$ coincide in this point.

From a dynamical point of view it seems natural to weaken the regular condition by assuming it to be satisfied only on an open dense subset. The first problem is to define a "good" notion for the singularities of bihamiltonian systems in odd dimension and for the Veronese webs transverse to the axis $\mathcal{A}$. This has been done for bihamiltonian systems in even dimension [16]. It would be interesting (but certainly difficult) to examine the odd dimensional case.

## References

1. T. ASUKE, *On transversely flat conformal foliations with good measures*, Trans. Amer. Math. Soc. **348** (1996), 1939–1958

2. M. A. AKIVIS & A. M. SHELEKOV, *Geometry and algebra of multidimensionnal three-webs*, Mathematics and its Applications (Soviet Series), vol. 82, Kluwer, Dordrecht, 1982

3. D. FRIED, *Closed similarity manifolds*, Comment. Math. Helv. **55** (1980), 576–582

4. I. M. GELFAND & I. ZAKHAREVITCH, *Webs, Veronese curves and bihamiltonian systems*, J. Funct. Anal. **99** (1991), 150–178

5. E. GHYS, *Flots transversalement affines et tissus feuilletés*, Mém. Soc. Math. France **46** (1991), 123–150

6. C. GODBILLON, *Feuilletages. Etudes géométriques*, Progress in Mathematics, vol. 98, Birkhäuser, Basel, 1991

7. W. KLINGENBERG, *Riemannian geometry*, second edition, Studies in Mathematics, vol. 1, de Gruyter, Berlin, 1995

8. S. KOBAYASHI & K. NOMIZU, *Foundations of differential geometry*, Interscience Publishers, New York-London, 1963

9. B. KOSTANT, *Systèmes hamiltoniens complètement intégrables*, Université Paris Nord, 1984

10. F. MAGRI & C. MOROSI, *A geometrical characterization of integrable hamiltonian systems through the theory of Poisson-Nijenhuis manifolds*, Publ. Dept. Math. Milan, 1984

11. P. MOLINO, *Riemannian foliations*, Progress in Mathematics, vol. 73, Birkhäuser, Boston, 1988

12. T. NISHIMORI, *A note on the classification of non singular flows with transverse similarity structures*, Hokkaido Math. J. **21** (1992), 381–393

13. M. H. RIGAL, *Rigidité des feuilletages transversalement conformément parallélisables*, Tôhoku Math. J. (2) **50** (1998), no. 3, 407–418

14. M. H. RIGAL, *Systèmes bihamiltoniens en dimension impaire*, Ann. Sci. École Norm. Sup. (4) **31** (1998), no. 3, 345–359

15. H. J. SÜSSMANN, *A generalization of the closed subgroup theorem to quotient of arbitrary manifolds*, J. Diff. Geom. **10** (1975), 151–166

16. N'G. TIENZUNG, *Symplectic topology of integrable Hamiltonian systems*, Thesis, Université de Strasbourg, 1994

17. I. VAISMAN, *Lectures on the geometry of Poisson manifolds*, Progress in Mathematics, vol. 118, Birkhaüser, Basel, 1994

# A THREE-DIMENSIONAL LAGRANGIAN FOUR-WEB WITH NO ABELIAN RELATION

GILLES F. ROBERT

*Laboratoire de Mathématiques pures, Université Bordeaux I et C.N.R.S., F-33405 Talence Cedex, robert@math.u-bordeaux.fr*

## 1  Introduction

Symplectic geometry is a field of choice for examples of webs: let $\mathbb{R}^{2n}$ be equipped with its canonical symplectic form $\omega = \sum dp_i \wedge dq_i$ and let $\Sigma \subset \mathbb{R}^{2n}$ be a Lagrangian sub-manifold. The 1-form $\alpha = \sum p_i \, dq_i$ defines a foliation on $\Sigma$ which, once projected onto $\mathbb{R}^n$, gives rise to a web on the space of positions $q_i$, provided that the projection has finite degree.

We now consider the Hamiltonian $\tau - \sqrt{\xi^2 + \eta^2}$ in $\mathbb{R}^6$ equipped with the symplectic form $d\xi \wedge dx + d\eta \wedge dy + d\tau \wedge dt$. We try to understand which Lagrangian sub-manifolds associated with this Hamiltonian give rise to webs having Abelian relations. Let $\Sigma$ be a Lagrangian sub-manifold of $\mathbb{R}^6$ associated with the Hamiltonian $\tau - \sqrt{\xi^2 + \eta^2}$, and let $u$ and $v$ be

$$u = x + t\frac{\xi}{\tau} \quad \text{and} \quad v = y + t\frac{\eta}{\tau}.$$

With these new parameters, the contact 1-form is:

$$\xi dx + \eta dy + \tau dt = \xi du + \eta dv + (\tau^2 - \xi^2 - \eta^2)\frac{dt}{\tau}$$
$$- \frac{t}{\tau}(\xi d\xi + \eta d\eta) + t(\xi^2 + \eta^2)\frac{d\tau}{\tau^2}$$
$$= \xi du + \eta dv.$$

Since the sub-manifold is assumed to be Lagrangian, this 1-form is closed, and there is a function $\psi(u, v)$, defined locally, that gives a local parameterization of $\Sigma$:

$$\begin{pmatrix} x \\ y \end{pmatrix} = \begin{pmatrix} u \\ v \end{pmatrix} - t\frac{\nabla\psi(u, v)}{|\nabla\psi(u, v)|}, \qquad \begin{pmatrix} \xi \\ \eta \end{pmatrix} = \nabla\psi(u, v),$$
$$t = t, \qquad\qquad\qquad \tau = |\nabla\psi(u, v)|.$$

From here on we assume that this parameterization is global, which means that the Lagrangian space describes the time evolution of a single phase function $\psi$ given at an initial time.

The map $\Phi$ from $\mathbb{R}^3$ to $\mathbb{R}^3$ defining $(x, y, t)$ in terms of $(u, v, t)$ gathers all the information needed to reconstruct the whole Lagrangian sub-manifold $\Sigma$; this map is defined by

$$\begin{pmatrix} x \\ y \end{pmatrix} = \begin{pmatrix} u \\ v \end{pmatrix} - t \frac{\nabla \psi(u, v)}{|\nabla \psi(u, v)|}.$$

Thus, except when $(u, v, t)$ belongs to the singular locus of this map, we can identify the 1-forms in the variables $(u, v, t)$ with the 1-forms in the variables $(x, y, t)$. In particular the contact 1-form $d\psi(u, v)$ can be expressed as a 1-form in the variables $(x, y, t)$.

Now, if $(x, y, t)$ possesses several preimages $(u_k, v_k, t)$, then each one of these preimages will define a 1-form which we denote as $d\psi_k(x, y, t)$.

The phase $\psi$ is said to be resonant near $(x_0, y_0, t_0)$ if the web defined, near $(x_0, y_0, t_0)$, by the 1-forms $d\psi_k$ admits an Abelian relation, which means that there exists functions $\mu_k(x, y, t)$ defined in the neighborhood of $(x_0, y_0, t_0)$ and such that

$$\sum \mu_k(x, y, t) d\psi_k(x, y, t) = 0 \qquad d\mu_k(x, y, t) \wedge d\psi_k(x, y, t) = 0. \qquad (*)$$

When $\psi$ is an affine function, its differential $d\psi$ is constant and the map $\Phi$ is one-to-one. In this case, we do not have a web, but a linear foliation, and the question of the existence of Abelian relations is void.

Next comes the case when $\psi$ is a polynomial of degree 2. We assume that the homogeneous component of highest degree is non-degenerate, so that translation and rotation of the coordinate system gives a phase function of the form $\psi(u, v) = au^2 + bv^2 + c$. We can even assume that $c = 0$, since $\psi$ only occurs through its differential $d\psi$.

In the particular case of a circular phase function, where $a = b$, the vector $\nabla \psi(u, v)$ has the same direction as $(u, v)$, and so does $(x, y)$; thus the forms $d\psi_k$ are collinear. In that case also the question of the existence of Abelian relations is void.

**Theorem.** *Let $\psi(u, v) = \frac{1}{2}(au^2 + bv^2)$ be a quadratic non-degenerate phase function (i.e. $a \neq 0$, $b \neq 0$ and $a \neq b$), then the map $\Phi$ is algebraic of degree 4 and the four-web defined by the 1-forms $d\psi_k$ admits no Abelian relation.*

*Remarks.* The web defined above is not strictly speaking a 4-web, but rather some part of the 4-web of the theorem, related to the algebraic Hamiltonian $\tau^2 - (\xi^2 + \eta^2)$. Indeed, the evolution described through this Hamiltonian occurs in the future (where it describes the same evolution as $\tau - \sqrt{\xi^2 + \eta^2}$), but also in the past (in which case it is related to $\tau + \sqrt{\xi^2 + \eta^2}$); the web we are interested in consists only of the leaves occurring in the future.

The main reason why the complete web is easier to study is its algebraic structure, thus we can consider it as the real part of a complex 4-web. The theorem states that the whole complex web admits no Abelian relation; this implies that the web we consider has no Abelian relation either.

Since the web is a 4-web of codimension 1 in $\mathbb{R}^3$, general results (Chern-Griffiths) show that the space of Abelian relations has dimension at most 1.

Now, it is obvious that our problem is invariant under the transformation

$$a \longleftrightarrow b, \qquad u \longleftrightarrow v, \qquad x \longleftrightarrow y.$$

It follows that, if the web associated to particular values of $(a, b)$ admits an Abelian relation, then so does the web associated to the pair $(b, a)$.

*Sketch of the proof.*

To prove the result, we will proceed in two steps: in the first step, we will find explicit functions $\lambda_k(x, y, t)$ solving the dependence relation:

$$\sum \lambda_k(x, y, t) d\psi_k(x, y, t) = 0$$

Effectively, since the four 1-forms $d\psi_k$ live in a 3-dimensional space, then they must be dependent. On the other hand, since these 1-forms span the whole space, then every dependence relation can be deduced from that one, which means that every solution $\mu_k(x, y, t)$ satisfying $(*)$ has the form $\mu_k(x, y, t) = f(x, y, t)\lambda_k(x, y, t)$.

In the second step of the proof, we derive from the four equations $d\mu_k(x, y, t) \wedge d\psi_k(x, y, t) = 0$ a system of linear equations which the differential $df(x, y, t)$ satisfies. Then we prove that this system has only the trivial solution 0, which gives rise to a contradiction.

## 2 Definitions

Let $\theta$ be the angle between $\nabla\psi$ and the horizontal axis. We have the three following relations:

$$x = u - t\cos\theta, \qquad au\sin\theta = bv\cos\theta.$$
$$y = v - t\sin\theta,$$

Each value of $\theta$ gives a preimage $(u, v, t) = (x + t\cos\theta, y + t\sin\theta, t)$

of $(x, y, t)$, provided that the third equation be satisfied, which reads:

$$a(x + t \cos \theta) \sin \theta = b(y + t \sin \theta) \cos \theta$$
$$(a - b)t \cos \theta \sin \theta = by \cos \theta - ax \sin \theta$$
$$\sin \theta \cos \theta = \frac{by}{(a - b)t} \cos \theta - \frac{ax}{(a - b)t} \sin \theta.$$

This equation is an algebraic equation of degree 4 in the variable $Z = e^{i\theta}$. Let $Z_k(x, y, t) = e^{i\theta_k(x,y,t)}$ denote the roots of this equation. We are then entitled to write $(u_k, v_k, t)$ as functions of $\theta_k(x, y, t)$:

$$u_k(x, y, t) = x + t \cos \theta_k(x, y, t), \qquad v_k(x, y, t) = y + t \sin \theta_k(x, y, t).$$

For the sake of simplicity, we denote $\theta_k$ and $u_k$ omitting the variables $x$, $y$ and $t$. Differentiating the previous system thus gives:

$$du_k = dx + \cos \theta_k \, dt - t \sin \theta_k \, d\theta_k$$
$$dv_k = dy + \sin \theta_k \, dt + t \cos \theta_k \, d\theta_k$$

and so

$$\begin{aligned} d\psi_k &= au_k \, du_k + bv_k \, dv_k \\ &= a(x + t \cos \theta_k)(dx + \cos \theta_k \, dt - t \sin \theta_k \, d\theta_k) \\ &\quad + b(y + t \sin \theta_k)(dy + \sin \theta_k \, dt + t \cos \theta_k \, d\theta_k) \\ &= t\rho_k(\cos \theta_k \, dx + \sin \theta_k \, dy + dt) \end{aligned}$$

where

$$\rho_k(x, y, t) = a\left(1 + \frac{x}{t \cos \theta_k}\right) = b\left(1 + \frac{y}{t \sin \theta_k}\right).$$

## 3   Dependence relations

The four angles $\theta_k$ satisfy the equation:

$$\sin \theta_k \cos \theta_k = \frac{by}{(a - b)t} \cos \theta_k - \frac{ax}{(a - b)t} \sin \theta_k.$$

If we set $\xi = -\dfrac{ax}{(a - b)t}$ and $\eta = \dfrac{by}{(a - b)t}$, this becomes

$$\sin \theta_k \cos \theta_k = \xi \sin \theta_k + \eta \cos \theta_k.$$

Now, if $Z_k = \exp(i\theta_k)$ and $\omega = \xi + i\eta$, the $Z_k$ are the four solutions of $Z^2 - \overline{Z}^2 = 2\omega Z - 2\overline{\omega}\overline{Z}$ which is equivalent, since $\overline{Z} = 1/Z$, to

$$Z^4 - 2\omega Z^3 + 2\overline{\omega} Z - 1 = 0.$$

We deduce immediately that $\prod Z_k = -1$ and $\sum Z_k = 2\omega$, which gives as derivatives:

$$\sum d\theta_k = 0, \qquad \sum i\, e^{i\theta_k}\, d\theta_k = 2(d\xi + i\, d\eta).$$

Differentiating the first equation gives:

$$(\cos 2\theta_k - \xi \cos\theta_k + \eta \sin\theta_k)\, d\theta_k = \sin\theta_k\, d\xi + \cos\theta_k\, d\eta.$$

Now if we set $\lambda_k = \cos 2\theta_k - \xi \cos\theta_k + \eta \sin\theta_k$, we get $\lambda_k\, d\theta_k = \sin\theta_k\, d\xi + \cos\theta_k\, d\eta$, and then

$$0 = \sum \frac{\sin\theta_k}{\lambda_k}\, d\xi + \sum \frac{\cos\theta_k}{\lambda_k}\, d\eta$$

$$2(d\xi + i\, d\eta) = \left[ -\sum \frac{\sin^2\theta_k}{\lambda_k} + i\sum \frac{\sin\theta_k \cos\theta_k}{\lambda_k} \right] d\xi$$
$$+ \left[ -\sum \frac{\sin\theta_k \cos\theta_k}{\lambda_k} + i\sum \frac{\cos^2\theta_k}{\lambda_k} \right] d\eta.$$

The following equalities follow:

$$\sum \frac{\sin\theta_k}{\lambda_k} = 0 \qquad \sum \frac{\cos\theta_k}{\lambda_k} = 0 \qquad \sum \frac{\sin\theta_k \cos\theta_k}{\lambda_k} = 0$$

$$\sum \frac{\sin^2\theta_k}{\lambda_k} = -2 \qquad \sum \frac{\cos^2\theta_k}{\lambda_k} = 2 \qquad \sum \frac{1}{\lambda_k} = -2 + 2 = 0.$$

And, with $d\psi_k = t\rho_k(\cos\theta_k\, dx + \sin\theta_k\, dy + dt)$, this implies the following dependence relation:

$$\sum \frac{d\psi_k}{\lambda_k \rho_k} = 0.$$

## 4   Integrating factor

Since the four 1-forms $d\psi_k$ span $\mathbb{R}^3$, every dependence relation is necessarily of the form $\sum f\, d\psi_k / \lambda_k \rho_k = 0$. Finding an Abelian relation then boils down to finding a function $f$ such that the 1-form $f d\psi_k/\lambda_k \rho_k$ be exact for each $k$; this is equivalent to the existence of functions $\alpha_k$ satisfying

$$\frac{df}{f} = \frac{d(\lambda_k \rho_k)}{\lambda_k \rho_k} + \alpha_k \, d\psi_k.$$

Let me recall that the values of $\lambda_k$ and $\rho_k$ are:

$$\lambda_k = \cos 2\theta_k - \xi \cos \theta_k + \eta \sin \theta_k$$
$$\rho_k = a \left(1 + \frac{x}{t \cos \theta_k}\right) = b \left(1 + \frac{y}{t \sin \theta_k}\right)$$
$$= a + \frac{(b-a)\xi}{\cos \theta_k} = b + \frac{(a-b)\eta}{\sin \theta_k}.$$

Fortunately, since we have $\partial(\lambda_k \rho_k)/\partial t = 0$, we can decompose and obtain:

$$\frac{1}{f}\frac{\partial f}{\partial \xi} = \frac{1}{\lambda_k \rho_k}\frac{\partial(\lambda_k \rho_k)}{\partial \xi} + \alpha_k \frac{\partial \psi_k}{\partial \xi} \qquad \frac{1}{f}\frac{\partial f}{\partial t} = \alpha_k \frac{\partial \psi_k}{\partial t}$$
$$\frac{1}{f}\frac{\partial f}{\partial \eta} = \frac{1}{\lambda_k \rho_k}\frac{\partial(\lambda_k \rho_k)}{\partial \eta} + \alpha_k \frac{\partial \psi_k}{\partial \eta}$$

As $df/f$ does not depend on the value of $k$, the third equation gives $\alpha_k$ from $\partial \psi_k/\partial t$:

$$d\psi_k = t\rho_k(\cos \theta_k \, dx + \sin \theta_k \, dy + dt)$$
$$= \frac{t\rho_k}{ab}\left(b(b-a)\cos\theta_k \, d\xi + a(a-b)\sin\theta_k \, d\eta\right.$$
$$+ \left.\left(ab + b(b-a)\xi \cos\theta_k + a(a-b)\eta \sin\theta_k\right) dt\right)$$

hence

$$\beta_k \alpha_k \, d\psi_k = \frac{1}{f}\frac{\partial f}{\partial t}\left(b(b-a)\cos\theta_k \, d\xi + a(a-b)\sin\theta_k \, d\eta + \beta_k \, dt\right)$$

with

$$\beta_k = ab + b(b-a)\xi \cos\theta_k + a(a-b)\eta \sin\theta_k.$$

Multiplying with $\beta_k \lambda_k^2 \rho_k$ gives the following equations for the components in $d\xi$ and $d\eta$:

$$\beta_k \lambda_k \frac{\partial(\lambda_k \rho_k)}{\partial \xi} = \beta_k \lambda_k^2 \rho_k \cdot \frac{1}{f}\frac{\partial f}{\partial \xi} - b(b-a)\lambda_k^2 \rho_k \cos\theta_k \cdot \frac{1}{f}\frac{\partial f}{\partial t},$$

$$\beta_k \lambda_k \frac{\partial(\lambda_k \rho_k)}{\partial \eta} = \beta_k \lambda_k^2 \rho_k \cdot \frac{1}{f}\frac{\partial f}{\partial \eta} - a(a-b)\lambda_k^2 \rho_k \sin\theta_k \cdot \frac{1}{f}\frac{\partial f}{\partial t}.$$

These equations can be viewed as equalities in the 4-dimensional $\mathbb{C}(\xi,\eta)$ vector space $\mathbb{C}(\xi,\eta)[e^{i\theta_k}]$. This means that we have to determine whether two families of three vectors are dependent or not.

Since we have the basis $(\cos^2\theta_k, \sin^2\theta_k, \xi\cos\theta_k, \eta\sin\theta_k)$, we compute the coordinates of the vectors in this basis; these coordinates are polynomials in $\xi$ and $\eta$. The existence of a dependence relation expresses the fact that some determinants, which are polynomials in $\xi$ and $\eta$, are in fact zero. The remaining part of this text deals with the computation of these polynomials at $\xi = \eta = 0$, which is enough to prove the theorem in the general case.

## 5  Computing $\lambda_k^2 \rho_k$

Before starting the computation, some useful equalities:

$$\xi\sin\theta_k + \eta\cos\theta_k = \cos\theta_k \sin\theta_k,$$

$$\frac{\eta\cos\theta_k}{\sin\theta_k} = \cos\theta_k - \xi, \qquad \frac{\xi\sin\theta_k}{\cos\theta_k} = \sin\theta_k - \eta,$$

$$\frac{\eta\cos^2\theta_k}{\sin\theta_k} = \cos^2\theta_k - \xi\cos\theta_k, \qquad \frac{\xi\sin^2\theta_k}{\cos\theta_k} = \sin^2\theta_k - \eta\sin\theta_k,$$

$$\frac{\eta}{\sin\theta_k} = \cos^2\theta_k - \xi\cos\theta_k + \eta\sin\theta_k, \qquad \frac{\xi}{\cos\theta_k} = \sin^2\theta_k - \eta\sin\theta_k + \xi\cos\theta_k.$$

From the values of $\lambda_k$ and $\rho_k$:

$$\lambda_k = \cos 2\theta_k - \xi\cos\theta_k + \eta\sin\theta_k$$

$$\rho_k = a + \frac{(b-a)\xi}{\cos\theta_k} = b + \frac{(a-b)\eta}{\sin\theta_k}$$

it is easy to deduce

$$\lambda_k^2 = 1 - \sin^2 2\theta_k + \xi^2 \cos^2 \theta_k + \eta^2 \sin^2 \theta_k - 2\xi\eta \sin \theta_k \cos \theta_k$$
$$- 2\xi \cos \theta_k (1 - 2\sin^2 \theta_k) + 2\eta \sin \theta_k (2\cos^2 \theta_k - 1)$$
$$= 1 - 4(\xi \sin \theta_k + \eta \cos \theta_k)^2 + \xi^2 \cos^2 \theta_k + \eta^2 \sin^2 \theta_k$$
$$- 2\xi\eta(\xi \sin \theta_k + \eta \cos \theta_k) + 4(\xi \sin \theta_k + \eta \cos \theta_k)(\xi \sin \theta_k + \eta \cos \theta_k)$$
$$- 2\xi \cos \theta_k - 2\eta \sin \theta_k$$
$$= (1 + \xi^2)(\cos^2 \theta_k - 2\eta \sin \theta_k) + (1 + \eta^2)(\sin^2 \theta_k - 2\xi \cos \theta_k)$$

and then

$$\lambda_k^2 \rho_k = \left((1 + \xi^2) \cos^2 \theta_k - 2(1 + \eta^2)\xi \cos \theta_k\right)\left(a + \frac{(b-a)\xi}{\cos \theta_k}\right)$$
$$+ \left((1 + \eta^2) \sin^2 \theta_k - 2(1 + \xi^2)\eta \sin \theta_k\right)\left(b + \frac{(a-b)\eta}{\sin \theta_k}\right)$$
$$= a(1 + \xi^2) \cos^2 \theta_k + \left((b-a)(1 + \xi^2) - 2a(1 + \eta^2)\right)\xi \cos \theta_k$$
$$+ b(1 + \eta^2) \sin^2 \theta_k + \left((a-b)(1 + \eta^2) - 2b(1 + \xi^2)\right)\eta \sin \theta_k$$
$$- 2(b-a)(1 + \eta^2)\xi^2 - 2(a-b)(1 + \xi^2)\eta^2$$

$$= \left(a + \varepsilon_2\right) \cos^2 \theta_k + \left(b + \varepsilon_2\right) \sin^2 \theta_k$$
$$+ \left(b - 3a + \varepsilon_2\right)\xi \cos \theta_k + \left(a - 3b + \varepsilon_2\right)\eta \sin \theta_k.$$

where $\varepsilon_k$ stands for any polynomial of degree $k$ vanishing at $(\xi, \eta) = (0, 0)$.

## 6 Computation of $\dfrac{\beta_k \lambda_k^2 \rho_k \xi}{\cos \theta_k}$ and $\dfrac{\beta_k \lambda_k^2 \rho_k \eta}{\sin \theta_k}$

The value of $\beta_k$ is:

$$\beta_k = ab + b(b-a)\xi \cos \theta_k + a(a-b)\eta \sin \theta_k$$
$$= a\left(b + (a-b)\eta \sin \theta_k\right) + b(b-a)\xi \cos \theta_k$$
$$= b\left(a + (b-a)\xi \cos \theta_k\right) + a(a-b)\eta \sin \theta_k.$$

Thus we get

$$\beta_k \lambda_k^2 \rho_k \frac{\xi}{\cos \theta_k} = a \left( \frac{b\xi}{\cos \theta_k} + \frac{(a-b)\xi\eta \sin \theta_k}{\cos \theta_k} \right) \lambda_k^2 \rho_k + b(b-a)\xi^2 \lambda_k^2 \rho_k$$

$$\beta_k \lambda_k^2 \rho_k \frac{\eta}{\sin \theta_k} = b \left( \frac{a\eta}{\sin \theta_k} + \frac{(b-a)\xi\eta \cos \theta_k}{\sin \theta_k} \right) \lambda_k^2 \rho_k + a(a-b)\eta^2 \lambda_k^2 \rho_k .$$

We shall now only develop

$$\left( \frac{b\xi}{\cos \theta_k} + \frac{(a-b)\xi\eta \sin \theta_k}{\cos \theta_k} \right) \lambda_k^2 \rho_k$$

$$= \left( a + \varepsilon_2 \right) \left( b\xi \cos \theta_k + (a-b)\xi\eta \sin \theta_k \cos \theta_k \right)$$

$$+ \left( b + \varepsilon_2 \right) \left( b(\sin^2 \theta_k - \eta \sin \theta_k) \right.$$

$$\left. + (a-b)\eta(\sin \theta_k - \eta) - (a-b)\xi\eta \sin \theta_k \cos \theta_k \right)$$

$$+ \left( b - 3a + \varepsilon_2 \right) \left( b\xi^2 + (a-b)\xi^2 \eta \sin \theta_k \right)$$

$$+ \left( a - 3b + \varepsilon_2 \right) \left( b\eta(\sin \theta_k - \eta) + (a-b)\eta^2 (\sin^2 \theta_k - \eta \sin \theta_k) \right) .$$

And this gives, once we gather terms:

$$\beta_k \lambda_k^2 \rho_k \frac{\xi}{\cos \theta_k} = \varepsilon_4 + (b^2 + \varepsilon_4) \sin^2 \theta_k + (ab + \varepsilon_4)\xi \cos \theta_k$$

$$+ \left( b(a - 2b + a - 3b) + \varepsilon_4 \right)\eta \sin \theta_k$$

$$= \varepsilon_4 \cos^2 \theta_k + (b^2 + \varepsilon_4) \sin^2 \theta_k$$

$$+ (ab + \varepsilon_4)\xi \cos \theta_k + (b(2a - 5b) + \varepsilon_4)\eta \sin \theta_k.$$

## 7 Computation of $\lambda_k \, d(\lambda_k \rho_k)$

Let me recall anew the values of $\lambda_k$ and $\rho_k$:

$$\lambda_k = \cos 2\theta_k - \xi \cos \theta_k + \eta \sin \theta_k$$

$$\rho_k = a + \frac{(b-a)\xi}{\cos \theta_k} = b + \frac{(a-b)\eta}{\sin \theta_k}.$$

This gives

$$\lambda_k \rho_k = (\cos^2 \theta_k - \xi \cos \theta_k)\left(a + \frac{(b-a)\xi}{\cos \theta_k}\right)$$
$$+ (-\sin^2 \theta_k + \eta \sin \theta_k)\left(b + \frac{(a-b)\eta}{\sin \theta_k}\right)$$
$$= a\cos^2 \theta_k + (b-2a)\xi \cos \theta_k - (b-a)\xi^2$$
$$- b\sin^2 \theta_k - (a-2b)\eta \sin \theta_k + (a-b)\eta^2$$

and then

$$d(\lambda_k \rho_k) = \left((b-2a)\cos \theta_k - 2(b-a)\xi\right) d\xi + \left(-(a-2b)\sin \theta_k + 2(a-b)\eta\right) d\eta$$
$$- \left(2a\cos \theta_k \sin \theta_k + (b-2a)\xi \sin \theta_k\right.$$
$$\left. + 2b\sin \theta_k \cos \theta_k + (a-2b)\eta \cos \theta_k\right) d\theta_k$$
$$= \left((b-2a)\cos \theta_k - 2(b-a)\xi\right) d\xi + \left(-(a-2b)\sin \theta_k + 2(a-b)\eta\right) d\eta$$
$$- \left(3b\xi \sin \theta_k + 3a\eta \cos \theta_k\right) d\theta_k.$$

From there we deduce

$$\lambda_k d(\lambda_k \rho_k) = \left((b-2a) - \frac{2(b-a)\xi}{\cos \theta_k}\right) \lambda_k \cos \theta_k \, d\xi$$
$$+ \left(-(a-2b) + \frac{2(a-b)\eta}{\sin \theta_k}\right) \lambda_k \sin \theta_k \, d\eta$$
$$- \left(3b\xi \sin \theta_k + 3a\eta \cos \theta_k\right)(\sin \theta_k \, d\xi + \cos \theta_k \, d\eta).$$

The developing step-by-step goes:

$$\left((b - 2a) - \frac{2(b-a)\xi}{\cos\theta_k}\right)\lambda_k = (b - 2a)(\cos 2\theta_k - \xi\cos\theta_k + \eta\sin\theta_k)$$
$$- 2(b - a)(\xi\cos\theta_k - \sin^2\theta_k + \eta\sin\theta_k$$
$$- \xi^2 + \eta\sin\theta_k - \eta^2)$$
$$= (b - 2a + \varepsilon_2)\cos^2\theta_k + (b + \varepsilon_2)\sin^2\theta_k$$
$$+ (4a - 3b)\xi\cos\theta_k + (2a - 3b)\eta\sin\theta_k$$
$$\left(-(a - 2b) + \frac{2(a-b)\eta}{\sin\theta_k}\right)\lambda_k = (a - 2b + \varepsilon_2)\sin^2\theta_k + (a + \varepsilon_2)\cos^2\theta_k$$
$$+ (4b - 3a)\eta\sin\theta_k + (2a - 3b)\xi\cos\theta_k$$

and then, since

$$\left(3b\xi\sin\theta_k + 3a\eta\cos\theta_k\right)\frac{\sin\theta_k}{\cos\theta_k} = 3b(\sin^2\theta_k - \eta\sin\theta_k) + 3a\eta\sin\theta_k$$

$$\left(3b\xi\sin\theta_k + 3a\eta\cos\theta_k\right)\frac{\cos\theta_k}{\sin\theta_k} = 3b\xi\cos\theta_k + 3a(\cos^2\theta_k - \xi\cos\theta_k)$$

we get:

$$\lambda_k d(\lambda_k\rho_k) = \Big((b - 2a + \varepsilon_2)\cos^2\theta_k + (-2b + \varepsilon_2)\sin^2\theta_k$$
$$+ (4a - 3b)\xi\cos\theta_k - a\eta\sin\theta_k\Big)\cos\theta_k\,d\xi$$
$$+ \Big((a - 2b + \varepsilon_2)\sin^2\theta_k + (-2a + \varepsilon_2)\cos^2\theta_k$$
$$+ (4b - 3a)\eta\sin\theta_k - b\xi\cos\theta_k\Big)\sin\theta_k\,d\eta\,.$$

## 8  Computation of $\beta_k\lambda_k d(\lambda_k\rho_k)$

The value of $\beta_k$ is:

$$\beta_k = ab + b(b - a)\xi\cos\theta_k + a(a - b)\eta\sin\theta_k\,.$$

Then we have

$$\frac{\beta_k \lambda_k}{\cos \theta_k} \frac{\partial(\lambda_k \rho_k)}{\partial \xi} = ab\Big((b - 2a + \varepsilon_2) \cos^2 \theta_k + (-2b + \varepsilon_2) \sin^2 \theta_k$$
$$+ (4a - 3b)\xi \cos \theta_k - a\eta \sin \theta_k\Big)$$
$$+ b(b - a)\Big((b - 2a + \varepsilon_2)\xi \cos \theta_k + \varepsilon_2 \cos^2 \theta_k + \varepsilon_2 \sin \theta_k \cos \theta_k$$
$$+ (2a - 3b + \varepsilon_2)\xi \sin \theta_k (\xi \sin \theta_k + \eta \cos \theta_k)\Big)$$
$$+ a(a - b)\Big((-2b + \varepsilon_2)\eta \sin \theta_k + \varepsilon_2 \sin^2 \theta_k + \varepsilon_2 \sin \theta_k \cos \theta_k$$
$$+ (3b - 2a + \varepsilon_2)\eta \cos \theta_k (\xi \sin \theta_k + \eta \cos \theta_k)\Big).$$

Or, after gathering terms:

$$\frac{\beta_k \lambda_k}{\cos \theta_k} \frac{\partial(\lambda_k \rho_k)}{\partial \xi} = \Big(ab(b - 2a) + \varepsilon_2\Big) \cos^2 \theta_k + \Big(-2ab^2 + \varepsilon_2\Big) \sin^2 \theta_k$$
$$+ b\Big(a(4a - 3b) + (b - a)(b - 2a) + \varepsilon_2\Big)\xi \cos \theta_k$$
$$- ab\Big(a + 2(a - b) + \varepsilon_2\Big)\eta \sin \theta_k$$
$$= \Big(ab(b - 2a) + \varepsilon_2\Big) \cos^2 \theta_k + \Big(-2ab^2 + \varepsilon_2\Big) \sin^2 \theta_k$$
$$+ b\Big(6a^2 - 6ab + b^2 + \varepsilon_2\Big)\xi \cos \theta_k - ab\Big(3a - 2b + \varepsilon_2\Big)\eta \sin \theta_k.$$

## 9    Conclusion

The matrix formed by the components of the three vectors, taken at $(\xi, \eta) = (0, 0)$ is

$$\begin{pmatrix} a & 0 & ab(b - 2a) \\ b & b^2 & -2ab^2 \\ b - 3a & ab & b(6a^2 - 6ab + b^2) \\ a - 3b & b(2a - 5b) & -ab(3a - 2b) \end{pmatrix}.$$

The only linear combination of these three vectors for which both the first and second component vanish is $w/b - (b - 2a)u + v$, and this combination is:

$$\begin{pmatrix} a(b - 2a) & -a(b - 2a) & \\ -2ab & -b(b - 2a) & +b^2 \\ 6a^2 - 6ab + b^2 - (b - 2a)(b - 3a) & +ab \\ -a(3a - 2b) & -(b - 2a)(a - 3b) + b(2a - 5b) \end{pmatrix} = \begin{pmatrix} 0 \\ 0 \\ 0 \\ -a^2 - 3ab - 2b^2 \end{pmatrix}.$$

Therefore, an Abelian relation cannot exist, except if $a$ and $b$ satisfy

$$a^2 + 3ab + 2b^2 = (a+b)(a+2b) = 0.$$

The solution $a = -2b$ must be rejected, since otherwise $b = -2a$ would also give an Abelian relation. There remains only the case $a + b = 0$.

In that case, we must carry on with the complete computation of the polynomials, and this gives for $(a,b) = (1,-1)$:

$$\lambda_k^2 \rho_k = \left(1 + 5\xi^2 - 4\eta^2\right) \cos^2 \theta_k + \left(-1 - 5\eta^2 + 4\xi^2\right) \sin^2 \theta_k$$
$$+ \left(-4 - 2\xi^2 - 2\eta^2\right) \xi \cos \theta_k + \left(4 + 2\eta^2 + 2\xi^2\right) \eta \sin \theta_k$$

$$\frac{\beta_k \lambda_k^2 \rho_k}{\cos \theta_k} = \left(4\xi^2 + 6\eta^2 + 2\xi^4 + 12\xi^2\eta^2 + 12\eta^4\right) \cos^2 \theta_k$$
$$+ \left(1 + 19\eta^2 + 2\xi^4 + 16\xi^2\eta^2 + 16\eta^4\right) \sin^2 \theta_k$$
$$+ \left(-1 - 5\xi^2 + 8\eta^2 + 2\xi^2\eta^2 + 2\eta^4\right) \xi \cos \theta_k$$
$$+ \left(-7 - 25\eta^2 + 6\xi^2 + 2\xi^4 - 4\xi^2\eta^2 - 4\eta^4\right) \eta \sin \theta_k$$

$$\frac{\beta_k \lambda_k}{\cos \theta_k} \frac{\partial(\lambda_k \rho_k)}{\partial \xi} = \left(3 + 18\xi^2 - 6\eta^2\right) \cos^2 \theta_k + \left(-2 + 14\xi^2 + 6\eta^2\right) \sin^2 \theta_k$$
$$+ \left(-13 - 8\xi^2 + 4\eta^2\right) \xi \cos \theta_k + \left(5 - 8\xi^2 - 8\eta^2\right) \eta \sin \theta_k \,.$$

The conclusion comes from the fact that the three vectors

$$\begin{pmatrix} 1 + 5\xi^2 - 4\eta^2 \\ -1 - 5\eta^2 + 4\xi^2 \\ -4 - 2\xi^2 - 2\eta^2 \\ 4 + 2\eta^2 + 2\xi^2 \end{pmatrix} \begin{pmatrix} 3 + 18\xi^2 - 6\eta^2 \\ -2 + 14\xi^2 + 6\eta^2 \\ -13 - 8\xi^2 + 4\eta^2 \\ 5 - 8\xi^2 - 8\eta^2 \end{pmatrix} \begin{pmatrix} 4\xi^2 + 6\eta^2 + 2\xi^4 + 12\xi^2\eta^2 + 12\eta^4 \\ 1 + 19\eta^2 + 2\xi^4 + 16\xi^2\eta^2 + 16\eta^4 \\ -1 - 5\xi^2 + 8\eta^2 + 2\xi^2\eta^2 + 2\eta^4 \\ -7 - 25\eta^2 + 6\xi^2 + 2\xi^4 - 4\xi^2\eta^2 - 4\eta^4 \end{pmatrix}$$

form an independent family in $\mathbb{C}(\xi, \eta)^4$.